Exploring Research Impact in Academia and Why It Matters

Posing fundamental questions around the worth of knowledge creation and the social value of in-depth research, this volume offers a novel approach by exploring why impact is important in academic research, rather than explaining how it should be conducted.

Using qualitative data to unpack what research impact really constitutes, this book foregrounds the practicalities of achieving impactful, high-quality academic research, and argues for the importance of best practice in instilling public and reputational value of research for wider societal gain. Chapters unpack the concept of impact and discuss how it can be made more tangible and realisable, particularly in the context of theoretical or pure research where research outcomes are often obscure. Calling for greater clarity in how to articulate the value of impact within research strategies, the book will ultimately argue for the central role of impact in core research processes and support the development of career researchers in their practical roles and identity formation.

The book will be of interest to academics and postgraduate students involved with research methods, research policy, and higher education more broadly. Despite the predominantly UK-based context of the research, the volume will have resonance in countries where knowledge economy concepts have impacted higher education policy and practice, and so research managers and higher education policy advisors may also find the book of interest.

Andy Phippen is Professor of Digital Rights in the Department of Computing and Informatics at Bournemouth University, UK.

Louise Rutt is Senior Impact and Knowledge Exchange Manager in the Faculty of Arts, Humanities and Business at the University of Plymouth, UK.

Routledge Research in Higher Education

The Development of University Teaching Over Time
Pedagogical Approaches from 1800 to the Present
Tom O'Donoghue

The Layered Landscape of Higher Education
Capturing Curriculum, Diversity and Cultures of Learning in Australia
Edited by Margaret Kumar, Supriya Pattanayak and Nish Belford

Authority, Passion, and Subjected-Centered Teaching
A Christian Pedagogical Philosophy
Christopher J. Richmann

Experiential Learning and Community Partnerships for Sustainable Development
A Foundational Model for Climate Action
Mara Huber, Michael Jabot, and Christina Heath

Philosophical Adventures in African Higher Education
Cultivating Doctoral Encounters within Democratic Citizenship Education
Edited by Yusef Waghid

Developing Feedback Literacy for Academic Journal Peer Review
Narratives from Researchers in Education and Applied Linguistics
Edited by Sin Wang Chong and Aurora Lixinhao Gao

Exploring Research Impact in Academia and Why It Matters
Perspectives on the Public Good and the Role of Research in Society
Andy Phippen and Louise Rutt

For more information about this series, please visit: www.routledge.com/Routledge-Research-in-Higher-Education/book-series/RRHE

Exploring Research Impact in Academia and Why It Matters

Perspectives on the Public Good and the Role of Research in Society

Andy Phippen and Louise Rutt

Routledge
Taylor & Francis Group
LONDON AND NEW YORK

First published 2025
by Routledge
4 Park Square, Milton Park, Abingdon, Oxon OX14 4RN

and by Routledge
605 Third Avenue, New York, NY 10158

Routledge is an imprint of the Taylor & Francis Group, an informa business

British Library Cataloguing-in-Publication Data
A catalogue record for this book is available from the British Library

ISBN: 9781032590073 (hbk)
ISBN: 9781032595030 (pbk)
ISBN: 9781003454991 (ebk)

DOI: 10.4324/9781003454991

Typeset in Galliard
by Newgen Publishing UK

Contents

1 Misunderstanding Impact

Introduction

This book looks at research impact and its place in the broader concept of public good and liberal education drawing upon John Stuart Mill's inaugural address at St Andrews in 1867 (Mill 1867) as a starting point. It argues that attempts to institutionalise impactful research in research assessment frameworks, institutional approaches to impact, and efforts to extend the concepts of research impact beyond the research arena are still poorly understood and will generally be underpinned with "enthusiasts" rather than deep research policy, long-term strategy, and incentivised approaches to reflect the value of impactful research. Furthermore, we will argue that attempting to extract impact from its fundamental position as a function of research does little to develop a deep understanding of the need for research impact and, perhaps more importantly, why research impact is important. The book both explores impact as a concept, which can be drawn from the history of the concept of public good and liberal education in universities as a theoretical foundation, but also draws heavily upon our own involvement in developing high-quality impactful research, sometimes in conflict with institutional policy and support.

We will critically explore literature that illustrates the "enthusiast" model of impact (Schnitzler et al. 2016; Davidson, Mahendra and Nicholson 2022), where researchers engage in impact as a "value add" or because it's nice to do, rather than seeing it as an essential part of research activity. Further exploration of the state of the art will also examine the conflicting perspective, where policy, rather than belief in the public good, has caused the concept of impact to evolve in the higher education landscape (Williams and Grant 2018; Cuthill et al. 2014; Calhoun 2006). Empirically, we draw upon 15 years' experience immersed in developing impactful research and research policy to empower impact in institutions. We reflect on experiences from a variety of institutions to argue that until universities fully understand what impact is it will never become embedded in higher education culture. However, the perspective of the discussion is international in nature and reflects on global reach of research and how nations set higher education policy to reflect the public good, against

DOI: 10.4324/9781003454991-1

a backdrop of neoliberal levers and their influence upon research policy and practice and raise fundamental questions about why we do research (Olssen and Peters 2005).

What this book is not is a "how to" guide for doing research impact. There are plenty of excellent resources, both online[123] and print publications (Bayley 2023; McKenna and McKenna 2021; Reichard et al. 2020; Reed 2018), that explore process and approach to conducting impactful research and how to write it up for research assessment. However, in this book we argue that unless you know WHY you are doing impact, HOW becomes a process that you engage with more likely because you've been told to, rather than because you want to. There is often a brief outline of the "why" in many existing texts on impact; Bayley suggests this is threefold: we must do impact (i.e. funding and assessment mandates it); we should do impact (external but change-driven drivers such as the Sustainable Development Goals (SDGs)); and we/I want to do impact (driven by personal motivation) (Bayley 2023: 25). It is the tensions between these motivations and the fact each drives a different kind of response to the impact agenda that we would argue is problematic for research impact in the United Kingdom. As Kelly (2022: xviii) says of those whose research is deeply transformative for others, "we need more of you".

Therefore, this book presents a fundamental question – why should research impact matter? The motivation for this is twofold.

First, we have been involved in both higher education and research impact for a considerable amount of time (see later), before (Andy) or right at the beginning (Louise) of when it became a defined part of research assessment exercises (Hicks 2012), and as such we have seen "impact culture", for want of a better word, grow in that time across the sector, driven, arguably, at a senior level in universities by a policy expectation that this should be part of the research process, and because it brings performance-based research funding with it. And second, over this period of time, we have both conducted our own work[4] and worked across the sector giving advice and guidance around how to conduct impactful research and how it is reported in assessment exercises. And over this period what has become clear to us is that research impact is still poorly understood, sometimes mistrusted, sometimes dismissed as not part of research, and in a lot of cases implemented purely because of the demands of the upcoming assessment exercise. When discussing whether there was a need for a book such as this, we could not escape from the fact that while research impact has been part of research assessment for over 10 years now in the United Kingdom, we still meet many in the sector who do not really understand it or, perhaps more importantly, why to do it.

After the UK Research Excellence Framework (REF2021),[5] we experienced both a concern and bewilderment that even after impact has been on the research policy agenda for well over 10 years (in terms of policy related to assessment), there seems to be very poor understanding of what it is, and how to do it in a way that is intrinsic to research. Once results were published, with vindication of our approach being a correct one, there seemed to be a frustration from those academics whose case studies had done well, and disinterest from research leaders in starting to build a holistic form of research and impact strategy within institutions because of the next REF being unknown.

Certainly, there seemed, in institutions with whom we both worked and knew of, that impact success in REF2021 resulted in, on the whole, a greater focus on impact support and development or prioritisation of those with impact potential for the future. Yet we also saw little evidence of impact being embedded in the culture of institutions or their academics, and there remained a prevailing view the impact was something we have to do, rather than something we want to do. And it was viewed as something that could, in general, be conducted by process and external experts, rather than embedded in the research culture of the institution.

For example, it was interesting in one university with whom we worked that after the REF2021 return had been submitted, the impact managers, who had done an excellent job in managing the process of supporting academics in writing case studies and chasing all of the underpinning evidence, were made redundant. A clear message from management in this institution that impact was something they need worry about only then the next REF is due, rather than something that forms a fundamental part of research management.

Two years later, impact manager posts were advertised, and those who had remained in the university and obtained different roles returned to their previous impact-based roles.

We can also reflect on Andy's path to impact in this brief reflection, developed in far more detail in Chapter 2, which further illustrates the ad hoc nature of how some institutions have engaged with impact.

> I began my academic career in a fairly standard computer science role looking at large network systems and implementing new systems and protocols to embrace the emergence of online technology. During this period, working with industry partners, I was asked to be involved in a project looking into trust and engagement with IT systems (Lacohée, Phippen and Furnell 2006). The work was driven by engaging with stakeholders around trust and privacy online, including citizens, civil society and policymakers. There was a view that young people (future citizens) were an important group to speak to and equally significant concerns around online harms arising among adults with caring responsibilities and in the media. However, our conversations with young people did not reflect this, they firstly understood the risks and, more significantly, could mitigate those risks. This work resulted (probably as a result in the significant efforts of comms and policy teams with the industry partners) in significant policy and media interviews, which then led to invitations to talk about this work in school settings and with non-governmental organisations who worked in the education sector.
>
> Personally, this was the first time I recognised the importance of feeding findings back into practice (via schools) and policy (via NGOs and policymakers). Early impact resulted in discussions with government and

policymakers, public talks and media interest and resulted in changes of practice in schools that had tangible benefit to students.

This was around 2005 and at the time this interest was not reflected by managers at my institution. The looming Research Assessment Exercise was research output focussed and they could not see any value in reaching beyond traditional academic publishing. "Focus on your outputs", I was told, "don't waste your time on other stuff".

Nevertheless, perhaps through sheer bloody mindedness, I continued on this path – working in the policy area meant positive outcomes for young people and education professionals and working with NGOs gave reach and access to data sets that would have been impossible working in a more insular academic mode, which led to higher quality and more significant outputs.

In 2011, the rebranded Research Excellence Framework announced that the 2014 process will consider "research impact". It was made clear impact would be part of the evaluation, and therefore subsequent performance related funding:

The assessment of impact will be based on expert review of case studies submitted by higher education institutions. Case studies may include any social, economic or cultural impact or benefit beyond academia that has taken place during the assessment period, underpinned by excellent research produced by the submitting institution within a given timeframe. Submissions will also include information about how the unit has supported and enabled impact during the assessment period.

Not long after this announced I bumped into my head of school in the corridor: "Hey Andy, have you seen this stuff about impact in the next REF? Don't you do this sort of thing?"

Suddenly, there was an interest, although there was not an understanding. As, therefore, the institution rushed in collecting of information, academics were told to "do more" in terms of being able to evidence impact, but there was a lack of support collecting this evidence, and poorly produced case studies were rushed and went through few iterations.I think I saw one draft of my case study before it was handed to the research and innovation department, and then a few months later I was told it had been submitted.

Surprisingly, when the results of REF2014 were published, outcomes were better than expected and there was, for a time, a move to try to understand impact more deeply. However, once the afterglow of the REF2014 success had died down, the interest waned once more. There was little management interest in developing a wider knowledge about impact as part of the research process. Outputs were what mattered.

We then fast forward to 2018 and meeting with newly appointed research manager who also took on responsibility for impact within my faculty (as a result of impacts growing importance for REF2020/2021), who also happens to be the co-author of this monograph. I was asked by them:

"You've got really good impact, why doesn't anyone seem to know about it?"

I responded:

"They've never seemed that interested or understand it, but I find it interesting and rewarding so I've just got on with it"

Working with someone who understood research policy and impact to a deep level was revelatory. We worked in a more methodical approach to building a case study (see Chapter 2), collecting evidence and getting some support. The case study was written early, checked, revised and refined.

However, at a departmental level there was still little interest. Far more focus was placed on outputs and, bizarrely, income (which as a single metric counts for an element of research environment and has the lowest bearing on REF outcomes), which were viewed as far more important. Perhaps because it reflected the focus of the head of department's own research career, which, if anything, shied away from trying to engage their work with external stakeholders.

Institutionally, the importance of impact was being realised, and there was more interest. Senior leaders started telling me what to do (even though they had no track record of doing this sort of work) and other research managers started telling me they needed to look at my "evidence" and what I need to do. Systems were put in place to "collect" evidence centrally. I made no use of those systems.

Nevertheless, the Unit of Assessment lead and the Deputy Vice Chancellor, one of the most inspirational research leaders I have ever worked with, were both supportive and I was "allowed" to work with this book's co-author without further interference because there was a view we knew what we were doing.

Yet institutionally impact knowledge was not being disseminated – academics in general didn't know what impact was or how to do it in general. Where pockets of impact excellence were being carried out, it was generally done by some enthusiastic academics who believed it was important, rather than through any institutional support. Institutionally there was a reliance on external knowledge and consultants were brought in. This resulted in the somewhat bizarre situation that I would be giving talks on impact and policy engagement at other institutions while my own bought in training from external partners.

Ultimately, we achieved an extremely good outcome in the REF for our unit of assessment. However, by then I had moved institutions. And experienced exactly the same environment. At the time of writing impact has been part of research assessment for 13 years, and in the next Research Excellence Framework it will count for 25% of performance-related funding. Yet it still seems to be being done on a wing and a prayer in many institutions.

This mirrors Louise's experience of impact for the first REF in a different institution:

> At the time of the first REF (REF2014) I was working as a REF Officer in a Social Sciences, Arts and Humanities Faculty in the eighteen months leading up to the deadline for submission, having recently completed a PhD in Human Geography. As REF Officer, the main purpose of my role was to organise the output submission, and collate some of the information for the environment statements. During that time, about a year prior to the submission deadline, it was fairly casually mentioned that the University would need to submit these things called impact case studies, and would I mind having a look at what to do with them. A knowledge exchange manager had been given the task of writing them, but besides some initial conversations and looking at who might have work that could be submitted in the form of a case study, it seemed that not very much had been done, and there wasn't an especially clear idea of the work that needed to happen next.
>
> So after a quick overview of what was known to date, and the suggestion of my supportive manager to attend one or two sector-wide events about impact (such as the LSE event "The Future of Academic Impacts" – 4th December 2012), I went off and made contact with the academics listed, to find out more about their research and where this might have had an impact. These conversations were often more to find out about half ideas regarding previous events and engagements, as well as more significant projects, as a starting point to then track down where the impact might have ended up, after which followed significant amounts of time on the internet doing desk research, trying to track down agendas, lists of names and other kinds of documents to see what had happened, as well as having conversations with key external project stakeholders to both find out what the value for them had been, and looking at where the evidence of this might be or could be constructed. This was not because of a lack of trust, but simply because the academic themself often didn't know.
>
> They'd worked on particular projects or in particular roles, and they'd simply not had to think about the relationship to their research or what the outcomes were. These academics were often very generous with their time to help work these details out, while also being somewhat bemused. It's hard to think of a discipline where this was a competitive process of selection, it was simply that there was some kind of lead; they perhaps had a reputation for engagement, and particularly public engagement, in the institution - some had a collaborative project involving significant non-academic partners or was funded by key and influential stakeholders in the area, or they had a significant non-academic role that was adjacent to their role as a researcher, such as chairing or being a member of a committee at a national or international level. The ultimate goal was just to have the number of case studies required, developed in such a way as to showcase

> the benefit most fully, to fulfil all the eligibility criteria, and to have tracked down and evidenced as much of the impact as possible. So what is incredible then, is that there were these people who were doing really interesting, valuable and engaged research, not because there was a thing called impact, but because it was just what they did and saw as intrinsic to their role as a researcher.
>
> During my PhD I often found it hard to understand why there was no sense within my academic community that the research I was doing (into fundraising and its governance within international development charities) was relevant beyond academia. I was interviewing fundraisers and donors, and beyond my suggestion of sending them a report summarising my findings (for which there was no time, funding or support), the institution was only interested in how I was going to deliver a PhD which made a contribution to my discipline. So as someone who had questioned the purpose of my PhD, and found non-applied research hard to relate to, or was perhaps for those with a freedom to think loftier thoughts than me, working on impact for REF2014 was eye-opening as to the way in which these people had developed their research and impact in the same breath. They had developed their own way of doing this, through a sense of curiosity and being part of something bigger than their academic disciplines, to achieve some quite incredible outcomes. These impacts weren't always easy to articulate in the case study format, but the impact was always there; the research was always fundamental to the change as required by the REF documentation, and it was exciting for it to matter.

So, in this book we will explore this in more detail and try to answer these questions of what impact is and why does it matter. The first one, we should suggest, is already well supported in both policy documents and literature, and we do not intend to repeat in significant detail the many excellent books and outputs on this. But it is important to place this in context before we move on to what is the most pressing question to explore of why does impact matter?

In carrying out this exploration, there will be elements of this book that will be highly ethnographic in nature (as has already been illustrated), drawing upon our own experiences across the sector. At points in this book various vignettes and reflections will be presented that allow us to explore the context of our own experiences. This is not to say we think we have all of the answers, but we have observed a great deal of how not to do this sort of thing.

We will also make use of Freedom of Information (FOI) data (Birkinshaw 2010), collected from institutions across the sector, to assess the nature of practice around impact. The questions we asked were:

1. Do you provide training for academics about research impact? If so, what is the nature of this training?
2. Do you provide training for professional service staff about research impact? If so, what is the nature of this training?

3 How many full time equivalent (FTE) professional services staff do you have dedicated to research impact?
4 How much did you spend on external consultants for research impact for REF2021?
5 Do you have a policy for the selection and development of impact case studies? If so, please can you provide it?
6 Is research impact a separate category for workload allocation for academics with significant responsibility for research (SRR)?
7 Do you have a civic university agreement? If so, please can you provide it?

This is a particularly telling broad analysis of how policy at a national level is transformed into practice within institutions and provides a level of vindication that perhaps we were not wrong to think that the sector was not effective at engaging with the impact agenda, which then allowed us to ask why this might be the case.

A final, and perhaps most significant, part of the research is drawn from interviews with peers from both the academic and impact professional community. All of the academics we spoke to have had considerable success with impact, both, typically, in research assessment terms but also as a core aspect of their academic identity. Interviews were conducted with a broad range of senior leaders, academics, impact professionals (many of whom are in professional services roles in institutions), and also some external stakeholders related to research impact to explore their experiences and stakeholder perspectives.

Academics were selected because they had played a senior management role in research leadership (including two who served as panel members in REF2021), were "impactful" researchers who had done well in REF2021, or were academics who had lot of stakeholder reach with their research but had not submitted an impact case study in either the REF2014 or REF2021.

The interview approach in all cases was semi-structured and while generally specific to the interview subject (e.g. senior leader, panel member, early career researcher (ECR), stakeholder, etc.), each academic interview did consider:

- Personal views on impact and why they chose to do research that engaged with their stakeholder base.
- Support they had received (training, mentorship, professional services support, money, workload allocation, etc.) from their institution specifically for their impact work.
- Challenges of stakeholder engagement and framing their impact.

Perhaps the most significant of these interviews, described in-depth in Chapter 6, was a conversation with an extremely well-cited research professor at a Russell Group university who has worked for almost 30 years in the sector. They were always of the view that research impact was not something that they would be interested in because, when explored at their previous institution, there were given a poor explanation of what impact is, and it was something they viewed in conclusion as yet another administrative process with little interest or value to their work. However, we knew that this person had broad reach in their research discipline and significant impact outside of the academic domain. And, post interview, is now working to show the impact of their work in anticipation of submitting an excellent case study in the new Research Excellence Framework.

In this case, it seemed, it was not that they weren't doing impactful research, they just couldn't see research impact as part of their academic identity due, predominantly, to poor articulation of why research impact might be something that could identify with.

This will be explored in far more detail throughout this book, that will begin by exploring the evolution of research impact as part of the research agenda, from concepts of public good through to the, arguably, neoliberalisation of the higher education sector in recent years. We will then explore, through various aspects of empirical work, why impact is so misunderstood (and evidence that this is the case) and how this might be addressed. Our motivation is not because we think now impact is part of research assessment it is important to understand it to impact assessment outcomes. It is because we have seen the value of impactful research and do not see it as separate from the academic identity. Equally, however, we can appreciate why the former possibility is the view of many academics.

An interesting, unexpected consequence of this empirical activity was something of a sanity check for us. As discussed earlier, working together, through the previous REF and subsequently, we have often expressed confusion that institutions still struggle with impact. Many academics do not understand it or see its significance, and those who did impactful research seemed to do so in spite of their institutions, rather than because of them. That is not to say that we think all academics should be doing impact, but we also saw some who had potential to be doing excellent impactful work but had chosen not to. We always had a nagging doubt that maybe it was us who were wrong, and our ideology that impactful research can be part of academic identity was wrong and the disinterest was justified. In total, we conducted 40 interviews. By around the third or fourth we realised that our frustrations, confusions, and concerns were also shared by others who had taken an impact path.

What Is Research Impact – The Policy View

Research England (2022) defines research impact as "an effect on, change or benefit to the economy, society, culture, public policy or services, health, the environment or quality of life, beyond academia". This definition encompasses a wide range of potential impacts, emphasising that research should not only advance academic knowledge but also contribute positively to various aspects of society.

We can draw together the threads from policy documents (ibid.) and the literature around research impact (e.g. Donovan 2011; Penfield et al. 2014;

Greenhalgh et al. 2016; Reed et al. 2021) to attempt to define what research impact is, at least as defined as part of a research assessment framework. This is a generalised overview of what impact can be, but it is a useful starting point to our investigation. It suggests that impact can be applied and assessed in different disciplines in different ways, but there seems to be various types of impact broadly defined:

- Economic impact: This includes the creation of new companies, improving business performance, job creation, and contributions to economic growth through innovation and commercialisation of research findings.
- Policy impact: Research and evidence developed from findings can influence legislation, public policy, and government or organisational decision-making, leading to changes in laws, regulations, and policy frameworks.
- Health and well-being impact: Research in medical and health-related fields can lead to the development of new treatments, drugs, healthcare policies, or interventions that improve patient outcomes and public health.
- Technological impact: The development of new technologies or improvement of existing ones can lead to productivity gains and advancements across various sectors, including information technology, engineering, and manufacturing.
- Environmental impact: Research can inform environmental policy, contribute to conservation efforts, lead to the development of sustainable technologies, and improve understanding of climate change and its effects.
- Social impact: Studies in social sciences, education, and humanities can lead to better understanding of social issues, inform social interventions, and lead to improvements in social cohesion and quality of life.
- Cultural impact: Research in the arts and humanities can influence cultural understanding, preservation, and expression, and can lead to societal benefits through increased cultural capital.
- Educational impact: Research can impact teaching methods and curriculum development, leading to better educational outcomes and advancements in pedagogy.
- Capacity building: This involves enhancing the ability of individuals, communities, or organisations to use research findings to solve problems and make informed decisions.
- International development: Research can influence international development projects and aid efforts, contributing to poverty reduction, global health initiatives, and international collaboration and understanding.
- Public engagement: Engaging with the public to disseminate research findings and to encourage public dialogue on scientific and societal issues is another key impact, fostering a more informed society.

Put simply, there are lots of different types of impact, and the nature of impact for a given piece of research, or even a research discipline, will depend a great deal on the stakeholder space within a discipline.

A further question in terms of assessing research impact is how does one evidence the claims made about the research impact being submitted? Again, drawing upon both policy and impact literature, we can also see that, in part because of the broad range of types of impact, assessing research impact is also complex and diverse. However, there are several methods and indicators commonly used to assess impact:

- Quantitative metrics: These might include citation counts, which indicate academic impact through the number of times a research output is cited by other researchers. Patent counts and technology licensing agreements can measure the economic and technological impact of research.
- Qualitative assessments: These involve narrative descriptions and case studies that illustrate the research's influence on policy, practice, health outcomes, social change, or cultural shifts. Such narratives can provide context and detail that quantitative metrics cannot.
- Economic analysis: Methods like cost–benefit analysis, social return on investment (SROI), and economic impact studies can quantify the economic benefits of research, such as contributions to GDP, job creation, or savings to healthcare systems.
- Bibliometrics and altmetrics: Besides traditional citation analysis (bibliometrics), altmetrics track the attention and dissemination of research outputs across various platforms, including social media, policy documents, news media, and online reference managers. This can indicate public engagement and societal interest but it's arguably not an exhaustive or particularly accurate technique (Costas, Zahedi and Wouters 2015).
- Surveys and interviews: Gathering feedback directly from stakeholders – such as policymakers, practitioners, or the public – about the influence of research on their work or lives can provide valuable insights into its impact.
- Policy influence tracking: This includes analysing references to research in policy documents, guidelines, or legislation to assess the extent to which academic work informs policy-making.
- Awards and recognitions: Achievements and recognitions, such as prizes for research or its applications, can serve as indicators of the societal value and impact of research.
- Health and environmental impact assessments: In specific fields, measuring the impact may involve tracking changes in health outcomes, environmental quality, or conservation status as a result of research-informed interventions or policies.
- Educational impact: This can be assessed through the adoption of new pedagogies, curricula changes based on research findings, or improvements in educational outcomes.

Each of these methods comes with its own set of challenges and limitations. Quantitative metrics, for example, may not fully capture the nuanced and long-term impacts of research, while qualitative assessments can be subjective, time-consuming, and potentially difficult to generalise from (if generalisation is needed). Therefore, a combination of methods is often used to provide a more comprehensive understanding of research impact.

A final popular metric discussed at length in both policy and literature is "reach and significance", which are two key dimensions used to evaluate the impact of research. Reach refers to the extent and breadth of the influence that the research has had beyond the academic community. It considers how widely the research has been disseminated and utilised, including the diversity and number of beneficiaries or audiences that have been affected. For instance, reach might be demonstrated through evidence of research impacting multiple sectors, being adopted internationally, or influencing many organisations, practitioners, or members of the public.

On the other hand, significance refers to the intensity or depth of the influence that the research has had on individuals, communities, organisations, or policies. It assesses the importance or value of the change or benefit brought about by the research, including the magnitude of the change and the degree to which it has altered understanding, practices, or outcomes. Examples of significant impact might include research leading to substantial improvements in public health outcomes, major policy changes, significant economic benefits, or profound cultural shifts.

An impactful piece of research ideally demonstrates both high reach and high significance. For instance, research that is adopted across numerous countries and leads to significant policy changes would be highly rated in terms of impact. The combination of wide dissemination and profound influence ensures that the research makes a meaningful contribution to society on multiple levels.

However, a balance between reach and significance is often considered in assessments. Research might have limited reach but could still be highly impactful if the changes it brings about are profoundly significant for a particular community or sector. For example, ground-breaking research into a rare disease might not impact on high numbers of the population but for those few it does, it could potentially save their lives, so is unquestionably significant, even if reach is low. This dual focus ensures that the evaluation captures a comprehensive picture of the research's contribution to society.

A Unit of Impact – The Case Study

While this initial review is deliberately high level with a focus on policy definitions for research impact, it would not be an effective top-level review without considering what has, we would argue (and this is reinforced with our interviews and FOI work), become the policy focus of impact – the impact case study. Within the definitions of previous Research Excellence Frameworks, the means to articulate research impact has been the case study. For a given unit of assessment, there is an expectation that an institution will submit, alongside staff and outputs, a number of impact case studies in proportion with their overall submission. According to the REF2021 guidance, each Unit of Assessment (UoA) for an institution needed to submit at least two case studies for returns under 20 FTE staff, increasing by one with the addition of each 15 staff returned up to 110, then one more for every additional 50 FTE staff returned.[6]

Therefore, in order to do well in the REF return regarding impact, there was a huge amount of investment in how research impact was presented. This is not surprising, given that senior management motivations will lie in maximising funding return. However, this is, we would argue, a highly reductionist approach to how research impact is done.

The way impact was presented in REF2021 was highly prescriptive. The case study is defined with a structure that must include:[7]

- Summary (500 words): Begin by explaining the research context, including why the research was necessary and what problem it aimed to solve.
- Research description (500 words): Detail the research process, methodologies, and key findings.
- Impact description (750 words): Describe the specific changes or benefits that resulted from the research. Use clear examples and quantify the impact where possible.
- Evidence (up to 10 references): Provide robust evidence to support your claims, such as data, testimonials, and citations from external sources.

We can certainly see, from a research assessment point of view, that having this structure and prescribed, space-limited, approach is useful – assessors do not wish to be overwhelmed with narrative and detail. However, it also means that those focussing on delivering a good impact return in institutions will also focus not on the case study as the goal of impactful research, rather than simply the lens through which it can be presented for research assessment. Which, as we have observed and will discuss further later, results in academics in institutions being asked "do you have a case study?" not "how is your research impactful and how can we support it?" or being told "You don't need to worry about impact, we've got enough for this unit of assessment".

Nevertheless, unsurprisingly given the potential returns for presenting impact case studies in a manner that is appealing to assessors, there has been a great deal of work around how to write impact case studies (ibid.). A lot of this work performs historical analysis on previous REF submissions in order to present these research findings and conclusions. Estimating the number of articles published about writing impact case studies is challenging due to the broad scope and interdisciplinary nature of the topic. However, cursory searches on Google Scholar have the results in thousands.

Again, this is not surprising, given the interest in how to best write case studies to maximise return. But we would argue (and do throughout this book) that the focus on case studies as the driving force behind "doing" impact contributes to the reductionist perspective of research impact. The case study drive means that the importance of stakeholder relationships, long-term impact, aligning research goals with social need, and all of those things that, as we discover throughout this book, are important for those academics who see impact are core to their identity, are neglected. We have been told on several occasions, terms like "it's not about the research, it's about how you present it" and, given the prescriptive nature of the case study in research assessments, this position is not without merit but does, we would argue, fail to grasp the importance of aligning impact with academic identity and how, fundamentally, good impact comes from good research.

While the final policy statements for the next Research Excellent Framework (in 2029) has yet to be finalised, there is a steer towards a more holistic presentation of impact,[8] reflecting, they claim, a broader and more holistic approach to research assessment. Impact is expanded to include both engagement and impact, emphasising how research contributes to society and the economy, and there is less emphasis on impact case study load, with a minimum requirement of one case study per disciplinary submission. However, each UoA level submission will also need to produce an "engagement and impact" statement, which looks more holistically at engagement and impact in the discipline. This might result in a broader interest in impact across disciplines, rather than focussing upon a couple of impact "stars" and telling other academics not to bother, but this remains to be seen.

Opposition to Impact

As we have already discussed, impact, and its role in performance-based research funding systems (PBRFS), is not without significant criticism and when we explore the policy position on impact that is perhaps not a surprise – a prescriptive and structurally imposed approach to how individuals and institutions should articulate how their research (and their research ideology) has been impactful is very much in conflict with the academic identity (which we will explore in far more detail in Chapter 6). Assessment approaches, such as the Research Excellence Framework (REF) in the United Kingdom, have faced significant opposition from the academic community (e.g. Peters and Roberts 2008; Thomas et al. 2020) due to several negative consequences. Critics argue that these systems prioritise quantifiable outputs, such as publication counts and citation metrics, over qualitative aspects of research, potentially marginalising important but less easily measurable work (Hicks 2012; Watermeyer and Olssen 2016). This focus can skew research agendas towards safer topics, stifling innovative and high-risk research. Additionally, the administrative burden involved in preparing for PBRFS assessments is seen as a major drawback (Pinar and Horne 2022), diverting significant time and resources from actual research and teaching. The complexity and bureaucratic nature of these assessments often lead to an increase in compliance roles within universities, further detracting from core academic activities.

This literature also highlights other criticisms of PBRFS, such as potentially limiting academic freedom and creativity by pressuring researchers to align their work with favoured criteria and metrics. These systems can also exacerbate existing inequalities between institutions and researchers, as well-resourced universities and established researchers tend to perform better, attracting more funding and consolidating their advantages. Conversely, disadvantaged institutions and researchers may struggle to compete, widening the gap in research capacity and opportunities. The focus on performance metrics can encourage gaming behaviours, where activities are manipulated to score higher in assessments rather than genuinely improving research quality. Additionally, the pressure to perform well can increase stress and reduce morale among academics, fostering a culture of constant scrutiny and performance anxiety.

A comment by Professor Peter Higgs,[9] a Nobel Prize–winning academic whose work is fundamental in our understanding of the physics of the universe, perhaps illustrates this frustration perfectly:

> After I retired it was quite a long time before I went back to my department. I thought I was well out of it. It wasn't my way of doing things any more. Today I wouldn't get an academic job. It's as simple as that. I don't think I would be regarded as productive enough.

It is no surprise, therefore, to see a great deal of resistance to research impact by the academic community:

> Impact capture as an academic exercise and component of PRFS (Performance-based Research Funding Systems) is frequently under-mined in academic circles by a lack of consensus or mismatch between what academics and research funders and regulators respectively recognise and "count" as impact and the extent to which an impact agenda is perceived by the former to be incompatible with established modes of academic practice and convention.

The previous quote, taken from Watermeyer's excellent article critiquing the imposition of research impact as part of the UK's 2014 Research Excellence Framework (Watermeyer 2014), is a strong articulation of the opposition of academics to the concept of research impact. Other authors have developed these arguments (Smith et al. 2020) to critically examine the growing emphasis on "impact" in academic research, particularly in the United Kingdom. In these critiques the authors explore how the demand for measurable impact on society and the economy is shaping the behaviour, culture, and ethos of academic researchers and institutions.

They assess the influence of the REF in both the positive and negative terms, noting that while it can drive innovation and societal benefits, it also pressures researchers to prioritise impact over scholarly quality and can lead to ethical dilemmas. They put forward a comprehensive argument that the research impact agenda may lead to instrumentalism in research, where choices about what to study are driven by potential impacts rather than scientific curiosity or social need, and discuss how it affects the integrity of research.

We will unpick the literature further in the early chapters of this book. However, this tension might be explained as a view by many academics (and this was indeed reflected in our interviews) that impact is viewed as something that is forced upon them by policymakers and managers not because it is viewed as an important aspect of research, but because government has decided value for money has to be demonstrated in this neoliberalised higher education environment. We would propose that, given the interviews we have conducted and the significant positivity towards impact by several academics, often in spite of, rather than because of, their institution, there is another way to consider impact. In this approach, we propose that if impact can be reframed from a top-down imposition to something intrinsic to the academic identity (at least for some academics – a point we must stress and reiterate throughout this book) the undoubted benefits that can be reaped for impactful research might be achieved in a more harmonious way.

Approaching the Why

Therefore, in approaching the "Why" of research impact, we present the book in the following way.

First, we begin with an examination of Andy's impactful research on online harms and a deconstruction of the submitted case study for REF2021. This is presented not as a "how to" but an example of adopting a long-term approach to impact that eschews policy directions and senior management steers for a belief in the value of the impact of itself. The review showcases how integrating impact into the research process rather than treating it as an afterthought leads to significant societal benefits. The case study, contrasting the more strategically planned REF2021 submission with the rushed REF2014 effort, emphasises the importance of long-term stakeholder engagement and evidence-based policy and practice changes. It highlights the benefits of collaboration with schools, NGOs, and legal professionals in improving online safety for children and underscores that impactful research must be driven by a commitment to the stakeholder group, supported by robust evidence and ongoing stakeholder interaction, rather than something imposed on academic practice from above.

From this point we return to the macro environment and reflect upon the policy space over time. Chapter 3 explores the nature of research impact within the context of changing higher education policies, particularly focusing on the United Kingdom. It discusses the shift from traditional liberal and civic educational ideals towards a more neoliberal, market-driven approach, which has impacted the perception and integration of research impact into academic practice. The chapter reiterates the point that for research impact to be truly effective, it must be an intrinsic part of academic identity rather than an externally imposed requirement. It examines various policy frameworks such as the Civic University Commission, the Knowledge Exchange Framework (KEF), and the United Nations Sustainable Development Goals (SDGs), analysing their roles and effectiveness in promoting research impact and questions whether any of these top-down approaches have been effective. Additionally, it highlights the tensions between top-down policy impositions and the intrinsic motivations of academics, arguing that genuine engagement with research impact requires alignment with academic values and support for long-term, impactful research endeavours.

We then move on to explore the gap between policy and institutional practice. In Chapter 4, we examine the relationship between civic concepts and the current higher education research landscape in the United Kingdom, highlighting the challenges institutions face in integrating research impact into academic practice. It presents further argument that impactful research arises when academics understand their societal role and knowledge creation rather than merely complying with research assessment frameworks. The chapter critiques the neoliberal influence on higher education, noting the tensions between institutional policies and the intrinsic motivations of researchers. It explores the evolution of impact policies, particularly the inclusion of impact in the REF2014, and the subsequent institutional responses. Through empirical analysis, the chapter reveals inconsistencies in how institutions approach research impact, the reliance on external consultants, and the disconnect between research impact as part of academic identity and its perception as a management-driven metric. It underscores the need for a more coherent and supportive approach to fostering impactful research within the academic community.

In Chapter 5, the role of research assessment in encouraging universities to embed impact as a core principle is examined, with a focus on the UK's Research Excellence Framework (REF). The analysis includes feedback from REF panels, narratives from universities, and ethnographic evidence from impact professionals. It explores how competition among higher education institutions, driven by the REF, influences the types of impact highlighted and supported. The chapter highlights the challenges in institutionalising impact, including the gap between academic potential for impact and the available resources and support systems. It also addresses the discrepancies between impact as understood by institutions for REF submissions and the broader concept of impact culture as articulated by many impact professionals. The chapter concludes that while the REF has been successful in some respects, significant gaps remain in effectively embedding impact within academic practices and institutional strategies.

Developing the themes identified from empirical work in Chapter 5, Chapter 6 emphasises that impactful research arises from a genuine integration of impact into academic practice rather than as a response to external assessments like the REF. It critiques the management-driven approach to research impact by many institutions, highlighting the ethical and practical challenges faced by academics when impact is imposed as an institutional goal rather than a natural outcome of good research. Through interviews with impactful academics, the chapter reveals that successful impact often emerges organically through deep engagement with stakeholders and alignment with personal academic values. It calls for a cultural shift in higher education institutions to support and value impact as a fundamental aspect of research, arguing that this requires rethinking leadership strategies and fostering an environment where impact is recognised and nurtured from the grassroots level. The chapter concludes by advocating for a broader recognition of impact activities and a redefinition of academic success to enhance the relevance and utility of research in addressing societal needs.

Bringing these themes together, Chapter 7 develops further the need for impact to be part of academic identity to be effective, exploring the motivations and attitudes of academics towards impact within the context of higher education policies. It makes use Bourdieu's theoretical framework, including the concepts of field, habitus, and prestige, to analyse how academics perceive and engage with impact. The chapter argues that impactful research arises from a genuine integration of impact into academic practice, driven by intrinsic motivations and personal values, rather than as a response to external assessments or institutional demands. Through interviews with academics, it reveals that those who naturally incorporate impact into their work do so out of personal conviction and see it as an essential part of their identity. The chapter calls for a shift from top-down management-driven approaches to a more supportive environment that aligns with core academic values, ensuring that impact becomes a respected and integral part of scholarly practice.

Finally, in bringing the book to a close, Chapter 8 argues for a holistic view of research impact, mindful of academic identity, and placing the academic at the centre of an impact ecosystem. It examines the concept this research impact ecosystem, emphasising the importance of integrating impact into academic practice rather than treating it as a compliance exercise. Using Bourdieu's theoretical framework and Bronfenbrenner's ecological systems theory, the chapter brings the discussions to a close with a modelling of the multifaceted nature of research impact and the barriers to its effective implementation in higher education institutions. It highlights the inconsistencies in institutional approaches, the over-reliance on external consultants, and the need for a more supportive and collaborative environment that recognises the intrinsic motivations of academics. The chapter argues for a cultural shift within institutions to foster a deep understanding of impact, driven by engaged and motivated academics, and supported by clear policies and adequate training. Ultimately, it calls for embedding impact into academic identity and practice to achieve meaningful and sustainable societal benefits.

Notes

1 https://blogs.lse.ac.uk/impactofsocialsciences/about-the-lse-impact-blog/ [Accessed June 2024].
2 www.fasttrackimpact.com/blog/categories/research-impact-guides [Accessed June 2024].
3 https://research-impact-toolkit.co.uk/impact-case-studies/ [Accessed June 2024].
4 As a professor who has worked in the technology policy world for 25 years, and as a research professional who has worked across the sector for 15 years and been responsible for the support and writing of *many* impact case studies in this time.
5 https://results2021.ref.ac.uk/ [Accessed June 2024].
6 www.researchprofessionalnews.com/rr-news-uk-views-of-the-uk-2022-6-ref-2021-the-case-study-cliff-edge-returns/ [Accessed June 2024].
7 https://2021.ref.ac.uk/media/1447/ref-2019_01-guidance-on-submissions.pdf [Accessed June 2024].
8 https://repository.jisc.ac.uk/9148/1/research-excellence-framework-2028-initial-decisions-report.pdf [Accessed June 2024].
9 www.theguardian.com/science/2013/dec/06/peter-higgs-boson-academic-system [Accessed June 2024].

References

Bayley, J. (2023). *Creating meaningful impact: The essential guide to developing an impact-literate mindset.* Emerald Publishing Limited.

Birkinshaw, P. (2010). *Freedom of Information: The law, the practice and the ideal.* Cambridge University Press.

Calhoun, C. (2006). The university and the public good. *Thesis Eleven*, *84*(1), 7–43.

Costas, R., Zahedi, Z., & Wouters, P. (2015). Do "altmetrics" correlate with citations? Extensive comparison of altmetric indicators with citations from a multidisciplinary perspective. *Journal of the Association for Information Science and Technology*, *66*(10), 2003–2019.

Cuthill, M., O'Shea, E., Wilson, B., & Viljoen, P. (2014). Universities and the public good: A review of knowledge exchange policy and related university practice in Australia. *Australian Universities' Review*, *56*(2), 36–46.

Davidson, M. M., Mahendra, N., & Nicholson, N. (2022). Creating clinical research impact through social media: Five easy steps to get started. *Perspectives of the ASHA Special Interest Groups*, *7*(3), 669–678.

Donovan, C. (2011). State of the art in assessing research impact: Introduction to a special issue. *Research Evaluation*, *20*(3), 175–179.

Greenhalgh, T., Raftery, J., Hanney, S., & Glover, M. (2016). Research impact: A narrative review. *BMC Medicine*, *14*, 1–16.

Hicks, D. (2012). Performance-based university research funding systems. *Research Policy*, *41*(2), 251–261.

Kelly, W. (Ed.). (2022). *The impactful academic: Building a research career that makes a difference.* Emerald Publishing Limited.

Lacohée, H., Phippen, A. D., & Furnell, S. M. (2006). Risk and restitution: Assessing how users establish online trust. *Computers & Security*, *25*(7), 486–493.

McKenna, H. P., & McKenna, H. P. (2021). *Research impact: The what, why, when and how* (pp. 1–15). Springer International Publishing.

Mill, J. S. (1867). *Inaugural address: Delivered to the University of St. Andrews, February 1st, 1867.* Longmans, Green, Reader, and Dyer.

Olssen, M., & Peters, M. A. (2005). Neoliberalism, higher education and the knowledge economy: From the free market to knowledge capitalism. *Journal of Education Policy*, *20*(3), 313–345.

Penfield, T., Baker, M. J., Scoble, R., & Wykes, M. C. (2014). Assessment, evaluations, and definitions of research impact: A review. *Research Evaluation*, *23*(1), 21–32.

Peters, M. A., & Roberts, P. (2008). More than 'outputs': A critique of performance-based research funding. In Peters, M. A., & Roberts, P. (Eds.), *Neoliberalism, Higher Education and Research* (pp. 71–82). Brill.

Pinar, M., & Horne, T. J. (2022). Assessing research excellence: evaluating the Research Excellence Framework. *Research Evaluation*, *31*(2), 173–187.

Reed, M. S. (2018). *The research impact handbook* (2nd edition). Fast Track Impact.

Reed, M. S., Ferré, M., Martin-Ortega, J., Blanche, R., Lawford-Rolfe, R., Dallimer, M., & Holden, J. (2021). Evaluating impact from research: A methodological framework. *Research Policy*, *50*(4), 104147.

Reichard, B., Reed, M. S., Chubb, J., Hall, G., Jowett, L., Peart, A., & Whittle, A. (2020). Writing impact case studies: A comparative study of high-scoring and low-scoring case studies from REF2014. *Palgrave Communications*, *6*(1), 1–17.

Research England. (2022). How Research England supports research excellence. www.ukri.org/who-we-are/research-england/research-excellence/ref-impact/ [Accessed June 2024].

Schnitzler, K., Davies, N., Ross, F., & Harris, R. (2016). Using Twitter™ to drive research impact: A discussion of strategies, opportunities and challenges. *International Journal of Nursing Studies*, *59*, 15–26.

Smith, K. E., Bandola-Gill, J., Meer, N., Stewart, E., & Watermeyer, R. (2020). *The impact agenda: Controversies, consequences and challenges.* Policy Press.

Thomas, D. A., Nedeva, M., Tirado, M. M., & Jacob, M. (2020). Changing research on research evaluation: A critical literature review to revisit the agenda. *Research Evaluation*, *29*(3), 275–288.

Watermeyer, R. (2014). Issues in the articulation of 'impact': The responses of UK academics to 'impact' as a new measure of research assessment. *Studies in Higher Education*, *39*(2), 359–377.

Watermeyer, R., & Olssen, M. (2016). 'Excellence' and exclusion: The individual costs of institutional competitiveness. *Minerva*, *54*, 201–218.

Williams, K., & Grant, J. (2018). A comparative review of how the policy and procedures to assess research impact evolved in Australia and the UK. *Research Evaluation*, *27*(2), 93–105.

2 A Case Study in Long-Term Research Impact

Introduction

In this chapter, as a starting point to our exploration of why impact should matter, we explore the nature of the work around Andy's successful (in both REF and public good terms) impact work as an illustration of the nature of the impact. We should stress, this presents an approach to deep impactful research, not the only way to do it. However, there are some issues to draw out of reviewing the 2021 case study against the broad context of academic identity and how impact can become an intrinsic part of academic practice, and this is a vehicle to demonstrate this. While the presentation of this impact will draw out sections of the impact case study, the narrative will explore HOW the research got to the point of presentation and highlight the long-term and personally motivated approach to this work. This research was not driven by the case study structure or the need "to do a case study" for the REF, the case study is merely a lens on which to present the impact that had taken place anyway because it was an intrinsic part of the research process.

In this case, the focus around online harms presents some opportunities in that it attracts attention in both political and public discourse. However, it also presented some challenges in that, because of its high profile in public discourse, there is much policy and practice driven by opinion, rather than evidence, and part of the exploration of the nature of the case illustrates the perennial problem of research failing to impact – it simply reports the state of something rather than what needs to change. The heart of this case study lies in evidence-based policy and practice change, but also highlights the importance the symbiosis of impact and research during a long-term research practice, rather than assuming that, as a result of publishing some research, impact will emerge, or one begins to do research because they need to produce a case study for the REF/promotion. As will be explored, engagement with stakeholders has resulted in further research opportunities and access to data that would be extremely challenging to obtain with more traditional means. It shows how working together for common purpose can result in better outcomes across the stakeholder space.

DOI: 10.4324/9781003454991-2

The vignette in Chapter 1 is a broad-brush review of the somewhat meandering path that led Andy to his current position as someone who has been engaged in his research discipline for over 20 years, and using the concepts we will develop through the chapters in the book, we are better placed to explore this in terms of prestige and the field, from the personal perspective of an impactful academic. As such, the majority of this chapter is a discussion around Andy's last case study, drawing in part on the narrative provided (written by Louise and Andy), then adding further context in how the different aspects of engagement and impact (and significance) came about from a research critical mass, rather than a linear process of publishing a specific article that led to this impact, or a project that had impacts from dissemination activities or stakeholders as part of the project team.

The research and impact approach will be explored, which centred on changing policy and practice to be more aligned with victim need and improving victim support, using an evidence-based model to challenge current policy and practice. However, this is also an exploration of how impact can be viewed as a fundamental, and arguably essential, part of a research process, rather than something that has been determined post project completion, or as an add-on once publications are in place.

The Impact – Transforming Approaches to Online Safety: Embedding an Ethos of Empowerment and Support into Safeguarding Policy, Legislation, and Practice

The following is a detailed narrative around working up to the point of submitting the case study for REF2021. However, there is also some historical context that is worthwhile to state prior to getting underway properly on exploring the 2021 work, if only due to the illustrations of the differentiation between those who see impact as part of their personal research culture, and those who are charged with research assessment and a top-down view of working up impact.

Work related to the broad topic addressed in the case study explored in this chapter was also submitted for the REF2014, although this case study was focused more upon a specific aspect – teen sexting. On reflection, while the case study in 2014 was still well regarded, it was somewhat rushed and reflected the notion of an imposed impact agenda – it was written in the final six months of the REF cycle for that period and done so from a mind-set of "we've got to do case studies, you do this sort of thing don't you?". There was certainly no impact strategy for REF2014, and it is probably fair to say that there was little steer from senior management or staff who understood how best to articulate impact. While it was carried out with the best of intentions, it does reflect upon what happens with a policy imposition that is neither well-understood nor appreciated by those tasked with carrying it out. It also shows how research assessment can drive a separation of motivation by the professional and the academic. In the case of the development of case studies in 2014, there was certainly a laissez faire view of what to produce – essentially, we don't really know what we're doing but well give it a go. And from the perspective of the research professionals involved in this strategy, who can blame them? They did not really have much support from senior management and ultimately were not in charge or had any steer on the research. They could do their best at presenting the case studies, but if they do not do very well can suggest it's due to the research and impact lacking quality, rather than anything to do with the processes they adopted.

For the 2021 case study, the mood was different. There was some expectation by those in senior management (possibly as a result of the 2014 case study doing better than expected by those in charge of these decisions) that there might be some value in looking at the impact case study for the next (assumed at the time to be 2020) REF. However, post REF2014 results, the focus of senior management once again returned to different priorities and there was little interest in looking strategically at investment in impact or any broader investment in research culture. While those of us who had a research identity that had embedded impact would carry on with our own work regardless of the interest of the institution, this more generally was rarely more than appraisal reviews or asking when research funding might be coming into the university.

From a personal perspective, I continued with my work regardless of interest from my employer, this was something I had been doing from around 2004 and continued to build in momentum through the ensuing REF period. At the period around the time of the previous REF, there had been increased policy interest in what was broadly referred to at the time as online safety, and we (myself and the stakeholders with whom I worked the most significantly) were in a position to provide an evidence base to support *some* of the policy decisions being made at the time, with the potential to realistically expect better policy outcomes for children and young people across the country. For example, the expectation that school inspection in the United Kingdom (OFSTED 2023) would comprise questioning around online safety came into being in 2012, and the statutory safeguarding requirements for English schools (UK Government 2024) started calling for training and practice to consider online harms in 2016. Stakeholders with whom I worked (see further) were instrumental in what went into these policy changes, supported with data I was analysing at the time. As a result of these policy changes, the wider children's workforce (particularly the education sector) increasingly prioritised online safety, which then led to further opportunities to work with those stakeholders to improve practice in their settings; a wonderfully serendipitous example of how conducting research to change policy can then result in further opportunity to feed into further, more impactful, research.

Around 2018, the wider university was also starting to take the impending (now only two years away) REF more seriously and had started to bring in "mock REF" events and centralised systems to begin recording outputs and impact evidence. Of particular interest to me, as someone viewed as "an impactful academic", was the introduction of a centralised system where I was encouraged to submit evidence of impact so others could "assess" its quality and provide "feedback" on how to improve what was there. Personally, until the arrival of this book's co-author at my institution, the only feedback I ever received was "do more" or "I don't understand what you've done but it looks like its ok". Also, because of the nature of my work (and something I have written about in depth in terms of the emotional and moralistic responses to online harms, particularly with young people (e.g. Phippen 2016)) there were often conversations about what the individual thought about the work in personal terms, and provided their thoughts on what could be done to stop online harms occurring. While these conversations were always interesting, they were rarely about the development of the impact case study or what support there might be available, more personal views on my work. In fact, I think I was put in touch with a couple of schools as a result of these conversations and individuals asking if I could speak in their children's school, because of concern and/or interest, so a grass-roots practice was established, again, because I felt it was important and had value. The resultant data and means to publish outputs from this work were serendipitous.

Returning to the systems that were put in place by the institution, however, I didn't meet a single academic who viewed these centralised systems as useful or supportive of their endeavours, more as a tool for surveillance and control. So very few engaged with them. This, of course, placed academics and professional service staff in tension. Few could articulate why these centralised impact repositories were important and any questioning of this usually resulted in "well, university X uses it for their impact" or "well, <senior manager Y> has said we've got to do it like this". So rather than working in partnership with professional service staff, a lack of belief in the systems that were put in place, and the lack of support other than criticism, led to something of a stalemate in how impact case studies were being developed. Generally, the most motivational offering that was proffered was "a good impact case study might be good for promotion".

However, as discussed in Chapter 1, it was with the arrival of the new research manager into my new faculty in 2018 (who become both co-author of the impact case study and also this book) that triggered a step function change in interest in the case study work and how it might best be supported. Perhaps the most significant input into these developments of the case study was not financial support or a metric-based system, but that I could now talk to someone who both understood the work I was doing and provided a sounding board about how best to articulate the impact and what further work might be needed (e.g. testimonials or further evidence examples) in a constructive way, which unsurprisingly proved far more productive than "you've got to do this because this is the way we do it" or "we've got to do this for the REF".

It would seem, predictably, that working with someone with deep knowledge of research policy and how to relate your work to this was a far greater motivator than an institutional consensus of "we're doing this because we have to so let's get on with it".

I will return to working practice as we explore the case study in depth, however, which we will now begin to do. In presenting the case study, I will break down the different facets explored against the broader backdrop of the long-term research work it informed, and how stakeholder engagement during the research process can elicit far greater impact than post-result impact. In the following discussion, paragraphs taken directly from the case study are enclosed, to distinguish from the narrative alongside it.

Impact Summary

> Professor Phippen has mobilised findings from his research on how children and young people use digital technologies to significantly improve online safety for children throughout the United Kingdom and in other national contexts. Phippen has achieved this by embedding an ethos that prioritises the empowerment of children and young people through education and the establishment of positive support mechanisms within changes to national policy and statutory inspection, in changes to safeguarding practices in educational and other child-centred contexts and by directly engaging with an estimated 20,000+ children, 10,000+ members of the child workforce, and 5000+ parents. In addition, Phippen's research has underpinned the establishment of legislation and support for victims of so-called revenge porn. The recognition of revenge porn as a crime means that thousands who have been subject to this form of abuse have been able to access expert advice, have damaging images removed, and to seek justice.

With this case study we were attempting to articulate how evidence-based policy can result in better outcomes for young people, the goal of any stakeholder involved in child protection. However, as already discussed earlier, and certainly a major aspect of my research in recent years (Phippen and Street 2022), this is also an area where individuals who work across the children's workforce will bring personal opinion and emotion to situations they might need to address in their professional setting, and it is important to make sure that policy change, and therefore practice, is informed by evidence and not simply opinion. We also aimed to show that policy change can also impact on the grass roots of practice, that research of this nature does not have to specifically target a specific stakeholder group, and policy change can result in real change in outcome for young people.

In moving forward with this discussion, it is worthwhile to state, in broad terms, the nature of stakeholder engagement with this work. Over time working in the online safety space and seeing legislation and politicians come and go, I and a co-author have developed a stakeholder model for online harms based upon Bronfenbrenner's ecology of child development (Bronfenbrenner

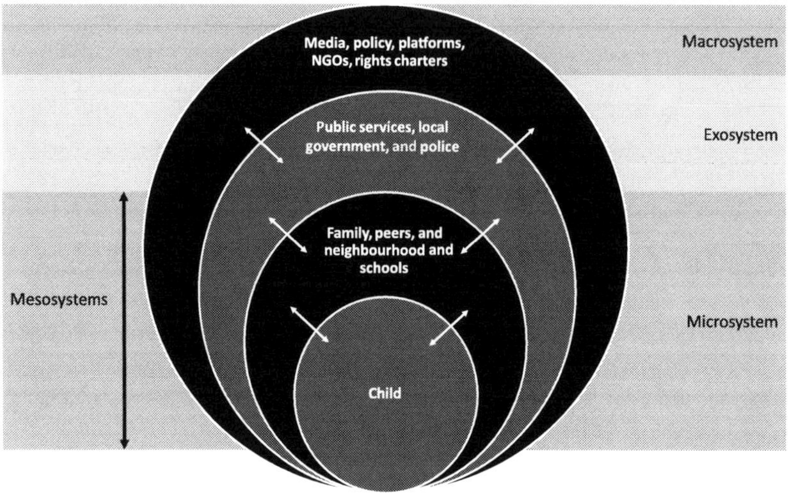

Figure 2.1 Online harms ecosystem.

1979), a well-established and widely used model for child development that can be developed to consider stakeholder responsibilities around online abuse. Bronfenbrenner's seminal work applied both nature and nurture perspectives to a child's development stating that their own biology is central to their development, but interactions with other actors in their environment, such as family, community, and society all play a part in development, and an action in one of these actors will impact across the wider ecosystem of development. Therefore, in order to provide the most effective context for child development we should not just look at the child but its environment as well. Consequently, given this extremely well-established model for child development, we can also use the model as a way of better understanding the ecosystem in keeping children safe online. Again, underpinned by research output, within this work, the following stakeholder model is increasingly viewed as a good summary of the ecosystem (Figure 2.1).

Developed from a theoretical underpinning of Bronfenbrenner's ecology of childhood (ibid.), and published in a number of outputs from this research (e.g. Phippen and Bond 2023), this helps us understand the nature of stakeholders around the child and their influence on them. Based upon the different systems defined in the model, they comprise the following in an online safety context:

- Microsystem: Family includes parents, siblings, and extended family, peers and neighbourhood, such as friends, fellow school pupils, and youth groups, and school and education settings, comprising teachers, support staff, school leaders, and the wider school community.

- Exosystem: Encompassing the broader children's workforce with safeguarding roles, such as police, social workers, statutory bodies, and healthcare professionals, who generally will only become involved in serious online harm incidents.
- Macrosystem: Which includes mass media, industry, policymakers, regulators, NGOs, and academia. These stakeholders are involved in national and international policy formation around online safety, which should influence the microsystems. The macrosystem will also be influenced by rights charters such as the UN Convention on the Rights of the Child, the European Convention on Human Rights, and the Universal Declaration of Human Rights.

The value of this model lies in highlighting the various stakeholders in online safeguarding and the importance of interactions (mesosystems) between those in the microsystem, as well the importance of communication between systems and the relative proximity of each stakeholder to the child. As Bronfenbrenner states, the quality of these interactions is crucial to positive outcomes for the child.

The ecosystem model also helps to take a holistic view of the stakeholder space, and helps to understand how best to influence each of the systems most effectively. As such, the primary stakeholders for this work were:

- **Schools and colleges:** Working with a few schools primarily, but also providing support for others who called for it, enabled a better understanding of the challenges those close to the safeguarding of children face, and how they interact with others around the child (both parents closer to the child and agencies present in other parts of the ecosystem). Schools also provided the opportunity to discuss issues with parents (e.g. in the event of being asked to give a talk to parents) and observe the dynamics between schools and parents.
- **Wider community (local authority):** As well as working in schools, part of the work was engaging with the wider children's workforce primarily through the Headstart Kernow project,[1] which is a broad education and social care project looking at early intervention mental health support across the county of Cornwall. A major strand of this work was a research project exploring the nature of online safety in regard to mental health support (Phippen and Street 2022) and provided opportunity to work with large numbers of young people and professionals. In addition, I worked with specific case work and delivered training opportunities which, in turn, would lead to a better understanding of this group's knowledge and attitudes and also of the challenges young people faced around online harms in the region.
- **Wider community (NGOs):** One of the primary stakeholders I have worked with over a long period of time is SWGfL,[2] a charity who have

worked in the online safety space for 20 years. Over this period of time, they have provided a wide range of products and services (including the Revenge Porn Helpline[3]) to help schools in their day-to-day educational and safeguarding practice tackling online harms. They have, over this time, become extremely influential among policymakers in how to develop effective national strategy around online harms, due to the access they have to the education sector and the knowledge they have developed through those using their services. I have worked with this charity for almost their entire life, and, as we will explore in more detail later, have been able to access significant data sets as a result. Furthermore, being able to work with this charity has provided the means to approach a national policy level that would have been very difficult to do working solely as an academic.
- **Legal community:** Working with Farrer and Co,[4] a law firm with a high-esteem safeguarding unit has also enabled the dissemination of research findings to a broad children's workforce audience (in particular through their Child on Child Abuse resource (Farrer and Co. 2023) but also through giving talks and networking with their community) as well as links at a national policy level with government departments.

However, one of the other benefits of this ecosystem model is that it allowed us to easily articulate the different levels of stakeholder engagement and why we were engaging with the stakeholder, so it also became a useful tool in articulating the impact process.

One question that is often raised to me when discussing impact either to support colleagues, or within broader gatherings around higher education policy, is how does one start to engage with stakeholders? I am sometimes told by junior colleagues who can see the value in doing impactful research and would like to know where to start with engaging with stakeholders with whom they might build relationships for their research. In terms of how to get started, there is already much discussion about how to identify stakeholders in your research discipline (e.g. Huzzard 2021), stakeholder mapping exercises (e.g. Reed, Bryce and Machen 2018), and similar, and I don't intend to replicate that here. However, one thing I did as part of this project was to speak to some of my own stakeholders, with many of whom I have had longer-term, mutually beneficial relationships over a significant number of years about their general engagement with academics. While we do not have the space to go into this in detail, there were a number of points raised that I was surprised about and helped better understand how to make a success of stakeholder engagement.

I will not mention stakeholders by name or organisation, but there were six people spoken to in total, drawn from different layers in the stakeholder model:

- A senior member of staff in an organisation from "wider community"
- A senior member of staff from "schools"

- A middle manager from "wider community"
- A policy manager from "industry"
- A director from "industry"
- A senior manager from "legal system"

What perhaps most surprising was there was general consensus about the themes that emerged. There was certainly no divergence of opinion from the different stakeholders.

Perhaps the strongest view from all stakeholders was that there was in their view a clear need for academics to engage more meaningfully with organisations for mutual benefit. Most interviewees had both bad and good experiences (explored further) but all saw the importance in academics engaging with stakeholders related to their research areas. This engagement was portrayed as beneficial for all parties, as academics can bring a systematic and scientific approach to product assessments and policy evaluations and provide evidence to work the stakeholder wishes to carry out or lobbying work. While some said that they could see the value in academic work that was more theoretical, most did say they could see some academics as being insular or disconnected from practical applications. It was important, in their view, that academics understand the social/industrial/economic/policy significance of what they did.

Another common theme that emerged a great deal was the importance of collaboration and long-term commitments. Interviewees in general said that, in their view, there was more benefit for both parties in having a long-term collaboration, and the fostering of ecosystems that include NGOs and other research groups can lead to more self-regulating and credible direction across the stakeholder space and a number also said that academics should also recognise the opportunity for stronger research outputs as a result of long-term collaboration. This, coincidentally, results in stronger research impact. However, a couple also warned against the potential toxicity of the relationship between academia and, for example, NGOs when it becomes too influenced by activism or lacks a balanced perspective.

Academics should be encouraged, in the view of interviewees, to bring their methodological rigour and skills in analysing data to industry and policy challenges. For example, in this area a number of stakeholders are concerned about emerging technologies such as artificial intelligence and virtual reality, but they wanted guidance from an independent evidence-based view as, as many stated, too much debate in this area is grounded in emotion and not evidence. However, the discussions also highlighted the mismatch between academic scales of assessment and industry practices, suggesting that more practical, informed, and systematic approaches are needed to align academic research with real-world needs. This is certainly a well-established tension between academia and its stakeholders, and something that one might view as being a driver behind policy drivers to "encourage" academics to engage in impactful research. Nevertheless, this doesn't mean that the stakeholders interviewed thought all academic research should be aligned with their needs, just that there is a lot of opportunity if there are those academics who view stakeholder engagement as a part of their research processes.

One stakeholder more explicitly brought this up given they had experience of working with academics from different geographies. They were of the view that those from a European school of thought were more likely to be rigorous than those they had dealt with from the United States (who they described as being more commercially/financially driven).

Perhaps most surprising for me with these interviews was that a number raised the issue of rigour and academic independence and how it is frequently compromised in their dealings with academics, where financial and political interests can potentially compromise research integrity. While this came as a surprise to me (probably because peers I worked with within my research discipline were from progressive and rational standpoints), perhaps, as we explore through this chapter, it is the nature of this ecosystem that emotion can inform debates too strongly. It might not be as apparent in other domains, but a number of interviewees talked about struggles with engaging with academics who would bring weak data and emotional bias to buy into popular policy and commercial perspectives (artificial intelligence was one area that was viewed as particularly problematic where academics would "pick a side" between stakeholders). There was a view that some academics would shape their research around personal opinions or emotional responses rather than solid evidence, which can dilute the objectivity and utility of their findings. A couple of interviewees also felt that some academic dissemination was driven by the pursuit of media attention rather than rigorous research. While there was some sympathy with this, given their understanding of (if not explicitly expressed this way) neoliberal leanings in the sector to be seen to be significant, there was definitely a view that this did not help them.

In summary, these stakeholders had a mature perspective on how and why academics can engage with stakeholders, but their views reflect ongoing challenges in bridging the gap between academic research and its practical application, advocating for a more engaged, evidence-based, and collaborative approach to academic work that remains grounded in rigorous methodology while being responsive to stakeholder needs and industry developments.

Underpinning Research

Returning to the case study, we now deconstruct the narrative around the underpinning research, we can see how engagement with a broad range of stakeholders, used in conjunction with mixed methods approaches to the research process, results in a broad understanding, and therefore evidence base, around online safeguarding. Note, the numerical references in this section refer to evidence sources, referred to later in this exploration of the case study.

The key aims of Prof. Phippen's research have been to establish an evidence base through which to understand how children and young people use digital technologies, and to critique safeguarding policy and legislation in this area. In so doing, he identifies how these tools of government and law could be made significantly more effective at protecting the safety of children and young people when they go online. Specifically, Phippen's research seeks to create evidence on children and young people's understanding of the opportunities and dangers afforded by the internet, of the relationship between online safeguarding and the consequences of this for their rights to use the internet, and the ways in which safeguarding policies and legislation play out for safety in practice. This is not a single project with rigidly defined aims, but an ongoing ethnography underpinned by the collection and analysis of large-scale quantitative survey data undertaken for charities who educate and advocate for children and young people. Phippen contextualises this with analysis of the policy and legal frameworks on interconnected but diverse topics such as teen sexting, revenge pornography, and young people's use of pornography. The ethnographic component draws on data from extensive engaged research with children and young people in schools (discussion groups, workshops, interviews), and talks and workshops with NGOs, the children's workforce, and those in the policy space. It is through this engaged approach to research that Phippen has been able to deliver impact directly to a large number of the children, young people, parents, and members of the children's workforce. Phippen uses these research interactions to share his research findings and analysis in a way that is relevant to participants.

As a result of this work being long term in nature, and also drawing upon many data sources and observational opportunities, it provided the means to see online harms from a broad perspective, rather than simply focusing upon one facet of online harms (as was the case with the 2014 case study). While there might be elements that do focus on a specific issue (e.g. sexting or revenge pornography), it is explored against a more holistic view of stakeholder knowledge and response. As such, with such a large number of evidence sources, one is in a better position to inform policy on the reality of what young people face, rather than only providing observations or evidence on a single phenomenon or edge case. Again, having such long-term relationships with stakeholders means that there is both excellent access to research data, and also dissemination paths outside of the traditional academic realm.

Specifically, Phippen (3.1) established the concept of an "online safeguarding dystopia" where, in a rush to apply adultist agendas from policymakers and children's workforce professionals to tackling online safety risks, children's rights are eroded and the idea of "safety" is an unachieved façade, rather than a meaningful reality. His research finds that this gap between legislation and practice has been brought about due to the development of policy without fair consideration of the evidence base, the capabilities of technology, or the rights of children and young people (3.1, 3.2, and 3.3), and is predominantly formed in the context of reactive lobbying. This has been coupled with a lack of education of the children's workforce (e.g. teachers, teaching assistants, social workers) about what legislation in this area means and how they can best support children and young people to be safer in their online behaviours (3.1, 3.3).

The main outcomes of these safeguarding failings have been twofold. First, they result in children and young people experiencing the safeguarding of their online activity as a form of soft-criminalisation, rather than as a supporting and trusting space (3.1, 3.2, 3.4, and 3.5). This means that rather than perceiving adults at school, home, out-of-school activities, etc. as supportive of them when they encounter problems or potential dangers, children and young people understand they will be punished for the consequences of anything that might be illegal (e.g. sexting and viewing pornography). Consequently, they often do not seek help when they need it (3.1, 3.2, and 3.3). Second, a dystopian vision of digital safeguarding at school and in the home places young people in a position where their behaviours and communications are filtered, monitored, and tracked. Not only does this equate to the significant neglect of children's rights but it also fails to teach them how to be safe online. It makes them either docile children who are compliant with restrictions without establishing their own understanding of how to manage and mitigate their own online risk, or devious children who hiding their potentially harmful online behaviours from caregivers, without an opportunity for education and intervention (3.1 and 3.3).

A lot of the observational work with stakeholders showed both an emotional response to a lot of child disclosure and an ambition from those professionals charged with safeguarding to prevent online harms from occurring. However, their knowledge was often poor, and they would frequently use legal threat to hope that this would dissuade young people from engaging with risky online behaviour (an approach we see in other areas of social policy such as drugs harms) (Nutt et al. 2007). We also observed many settings (e.g. schools) placing a reliance on technology to either prevent access to "harmful" material, or to detect abusive behaviour, and they would have policies in place which generally adopted prohibitive sanctions. Nevertheless, this contrasted with discussions with young people who raised concerns about the knowledge of those adults around them and a lack of willingness to provide support. This, young people said, was more of a controlling approach rather than helping them which, in turn, meant that they were far less likely to ask for help. So, as we showed with the research, this tension between those claiming to be safeguarding and those wanting support resulted in what I termed a dystopian response to safeguarding in this area, which we could evidence was neither what young people were asking for nor successful in its goals. The more recent longitudinal work across the stakeholder space in a particular region of England (Phippen and Street 2022), in particular, enabled first this dystopian practice to be demonstrated clearly and also that most policy (and research), which focused on measuring young people's behaviour and changing it, was far less effective than being able to provide a stakeholder ecosystem where young people were confident of disclosure and support, rather than being affected by online harms but scared to ask for help because of their fear of the consequences from ill-informed adults whose response would be emotional rather than constructive. By being able to evidence this we were able to take this to policymakers to call for greater statutory pressure for effective training in this area, as well to try to challenge policy views that technology can solve these issues without the intervention of other stakeholders (which remains ongoing work). Again, this work, while published in a number of traditional academic outputs in the first instance, was able to achieve far greater reach through working with stakeholders who could literally call up a government department to discuss this, than assuming or hoping that by some form of academic osmosis these findings would somehow reach those who might elicit change which resulted in better outcomes for young people.

Associated Research Outputs

The research outputs listed in this case study were:

3.1 Phippen, A. (2016). *Children's online behaviour and safety policy and rights challenges*. Palgrave.
3.2 Gray, S.L., and Phippen, A. (2017). *Invisibly blighted: The digital erosion of childhood*. UCL Institute of Education Press.
3.3 Phippen, A. and Brennan, M. (2019). *Child protection and safeguarding technologies*. Routledge.

These peer-reviewed outputs have been supplemented by peer reviewed practitioner reports, including:

3.4 Phippen, A. (2012). *Sexting: an exploration of practices, attitudes and influences*. NSPCC.
3.5 Bond, E. and Phippen, A. (2019). Police response to youth offending around the generation and distribution of indecent images of children and its implications. Research conducted for the Marie Collins Foundation.

Underpinning research was predominantly done on peer-reviewed books (given the volumes of data and associated narratives, as well as the importance in getting findings out quickly, books were always a better output type than other academic routes), but we also listed a couple of outputs done as peer-reviewed practitioner reports. The practitioner reports were illustrative, in order to highlight that such reports have a role to play in disseminating academically rigorous work. While the majority of the analysis from these reports ends up in some form of academic output, sometimes this is the quickest way to reach an audience about a pressing matter. For example, in the case of the Police Response to Youth Offending, there was a need to reach policymakers in understanding how young people were being treated very differently depending on where they lived in the country and the responding police professional. By producing a report for an NGO initially, we could disseminate it through their channels, and use their lobbying power to raise awareness. Some of this work did end up in a book (Phippen and Bond 2023) and the lobbying challenges in this area are ongoing, but this was not something we felt compelled to sit on until we could find a route for academic dissemination.

Furthermore, while these outputs were used to show the underpinning research, and a body of knowledge around this impact, the underpinning research for this case study is built on earlier work too, and the presentation of this underpinning research is more an illustration of the restrictions of the case study format for REF2021. There is a considerable body of knowledge behind the case study building upon work first published in 2006/2007 (Lacohée, Phippen and Furnell 2006).

Details of the Impact

Professor Phippen has significantly improved the safety of children and young people online throughout the United Kingdom and other national contexts through extensive in-depth stakeholder engagement at all levels of the impact pathway (direct engagement with children, capacity building with parents and the children's workforce, policymaker and public discourse, and government policy and legislation). Through these engagements, Phippen has articulated the findings from his research on how children and young people use digital technologies. This has led to the development of an established approach for safeguarding policies, regulation, and practices that is about creating empowering spaces for children and young people to go online. The methodology involves developing a knowledge of children and young people, and of those safeguarding them, about what safe online practices are, and providing the means to implement appropriate policies and support mechanisms in the current legal and technological context. Phippen has also translated these findings into other aspects of digital safety, leading to improved student support within higher education, and legal protections and support services for victims of so-called revenge porn.

Improving Safeguarding Practices and Cultures in Children's Spaces

Phippen has directly improved the safeguarding of children online by changing the way in which those with direct responsibility for this (school governing bodies, the children's workforce, and parents) understand and practice safeguarding. One way in which he has achieved this is through leading the commissioned SWGfL annual review of data from their 360-degree safe tool, every year since it was available in 2011. By 2020, this was well established as the leading self-review tool, used by 15,000 schools (responsible for an estimated 5 million children) in the United Kingdom. The availability of this data not only provides schools with the ability to benchmark against best practice but, as the director of SWGfL states, is

> used by SWGfL as a main source of data on current online safeguarding practice within schools nationally, and…has strongly influenced our strategy as an NGO, shaping the rest of our services and activities. This includes SWGfL addressing a weakness in training provision that Phippen identified in the data review and led to them developing their training offer to schools, Online Safety Live. This is a programme of e-safety events designed exclusively for professionals working with children and young people which, from 2015 to the end of 2020, have been delivered to over 22,000 professionals, collectively responsible for approximately 4.5 million young people.

This first aspect of the details of the impact is perhaps the strongest illustration of how working in partnership with an NGO for mutual benefit results in outcomes that are more powerful than either could achieve on their own. One of the services offered by SWGfL is a self-review tool for schools to reflect upon their online safety policy and practice.[5] Developed in 2009, as a result of both schools asking for it and the charity feeling that there was a need for it, this has been in place now for well over 10 years. Throughout this period of time, I have had access to the data collected by the tool (which is a simple database-driven website where all review ratings by a school are stored centrally) and have provided an annual "state of the nation" report (e.g. Phippen 2023) to see who was using the tool and what the state of online safety policy and practice in schools was. As an interesting aside, while early adoption of the tool was slow, with around 500 schools using it in its first year, the announcement by the school's regulator that they would inspect online safety policy and practice in 2012 resulted in a significant increase in the tool's use. Which, in turn, allowed us to demonstrate the impact of policy change on school practice. At the present time over 16,000 schools use the tool, and I still conduct the national analysis on this data, which shows factual information about how, for example, staff training on online safety is still one of the weakest aspects of practice in this area, even though there is a statutory duty to do so (which allows me to evidence a failure of policy that negatively impacts on the safeguarding of young people). If I was not working in partnership with this NGO who provide this service, being able to generate and use this data as an academic would be far more challenging. While not impossible, trying to contact and access information from 16,000 schools on their online safety policy and practice would be extremely difficult to do independently. And the reach would probably be far less significant too.

> Phippen has also been directly engaged in developing two training toolkits for the children's workforce (including teachers, social workers, healthcare professionals, and law enforcement officials). He wrote elements of the Farrer's Peer on Peer Abuse Toolkit in 2019 which was promoted in the DfE guidance on sexual violence and sexual harassment between children in schools and colleges for school and academy leaders (May 2018). The toolkit was also sent directly to approximately 3000 people on the Farrer's Safeguarding Unit's mailing list (spanning a broad range of sectors, including charity, sport, and religious organisations). In addition, he led and co-wrote the Headstart Cornwall commissioned Online Resilience Tool in 2019–2020. Phippen has also delivered training to an estimated 10,000+ members of the children's workforce and 5000+ parents between August 2013 and July 2020. The toolkits and training are designed to ensure that adults with safeguarding responsibilities (a) understand how children and young people use digital technology; (b) are equipped to support children and young people, particularly when they are exposed to risk and harm; and (c) are able to create an informed and non-judgemental culture within children's spaces (schools, sports, and other clubs, youth groups, etc.). In addition, they include practical actions to keep children safe (e.g. through reporting and the removal of harmful material). Evidence of practice change and the embedding of the principles is seen in the data from the 2020 SWGfL 360-degree annual review "which shows Whole School approaches being far stronger than they were ten years ago, and the vast majority of schools now [have] strong policies that underpin online safety practice in the schools".

In further partnership work that took research findings into practice, working with Farrer on their Peer on Peer Abuse toolkit (now referred to as the Child on Child Abuse toolkit (Farrer and Co. 2023)) was again a result of seeing the value of working together for common purpose. The law firm wanted to provide a free resource to their client base (and further afield), and I would imagine a part of that was promotion of their other services. However, it was something that was free to access, and there is no need to buy in other services to use it. From my perspective the relationship with Farrer resulted from early work in teen sexting (an aspect of child abuse) that one of their partners read and thought it might be useful to speak to me about. Given their reach, and the impact of this reach, not just with schools but with many organisations who work with children and young people meant that the more progressive messaging that young people had called for with that strand of research, was able to reach a large number of practitioners. As with the work with SWGfL, as a result of partnership working, we now have a long-established working relationship which has extended the reach and significance of my work and allowed me to explore a separate strand of the stakeholder ecosystem (legal services) in my own research.

The work with Headstart was strongly youth-led and resulted in the development of a resource for professionals to use in their safeguarding actions when supporting young people. From a very research-informed position we are able to get a more progressive approach into schools which is more aligned with what young people are calling for. As a result of the wider project, the tool was immediately disseminated to all schools in the county of Cornwall, 271 in total, reaching an estimated 100,000 young people. However, due to the manner in which we have developed and released it under Creative Commons licensing, the actual reach is far beyond the county. This research was eventually published in a monograph (Phippen and Street 2022), but it had significant reach prior to that publication as a result in the dissemination of the tool via local authority networks. Furthermore, given the relationships with schools established during the first round of research, we have been able to conduct a review of, and revisions to, the tool with young people who were involved in its initial development.

Empowering Children and Young People to Be Safe Online

In addition to changing safeguarding practices, Phippen also works directly with children to help them understand the risks of going online and how to respond to harms without this impacting on their rights and legislative protections. Between August 2013 and July 2020, Phippen has led online safety sessions with an estimated 20,000+ children (aged 4–18 years) in schools throughout the southwest and other parts of England to help them: to develop their critical thinking around online behaviours and risk; to understand their rights and how they can become resilient to harm; and to provide them with a forum to develop their knowledge around their online rights and responsibilities.

For about 10 years, from 2010 until the COVID lockdowns in 2020, I was often asked to go into schools to speak to young people, either in assemblies or in class, about online harms. Generally, I viewed this as a means to communicate a better understanding of online harms to young people and grow their confidence in asking for help. They also provided a lot of opportunities to discuss online harms and the online safety ecosystem with young people themselves to determine its efficacy and further inform research and practice. Given the long-term nature of this aspect of what I might refer to as public engagement rather than pure research, also provided a great deal of weight to policy arguments we were making – as we could authentically talk about this from the youth voice.

> For example, since opening in 2017, Phippen has worked with over 1000 young people aged 16–18 years at Callywith College, to develop their knowledge of digital rights and legislation. The head teacher writes "He discusses online safety with students, listening to their concerns and answering their questions, creating a space where they are able to be open and ask frank questions, knowing they will get an honest and supportive response". David Wright (director of SWGfL) also testifies to this in the foreword to Phippen (2016) "most importantly, [Phippen] speaks to children, a lot. And not just to deliver 'information' sessions, but sitting with them, working with them, answering their questions, and most importantly listening to them". Some Callywith students worked with Phippen to develop the Headstart toolkit and this subsequently led to them creating a film (July 2020) with the BBC, giving advice for staying safe online. He has also provided advice to students on a 1-to-1 level to help prevent them from risk, such as "advising and reassuring a student who was being blackmailed for money online, at risk to his reputation and future prospects. The student was able to deal with the incident with appropriate support from the Police and those responsible for his welfare, to achieve a positive outcome".

More recently I have worked a great deal with a specific college in Cornwall, as their college culture was progressive in nature which aligned strongly with our research findings, and as a result of our collaboration I have also become a trustee of the college. As illustrated in the previous narrative, this allowed a deep dive into practice in a specific location which might in the first instance be viewed as somewhat small scale compared with other aspects of the research. Establishing a long-term relationship with the college means that since its opening, alongside college growth as a result of their success, the research-informed approach to safeguarding they have developed has impacted upon well over 10,000 students. At present this relationship continues and we have implemented changes in pastoral support at the college that would have, again, been impossible without my relationship with the organisation. There is no way as an academic I could have walked into the place and suggested they change things without developing our mutual association first.

Improving child online safety through changing national policy, regulation, and shaping national and international discourse, Phippen has drawn on the evidence base from his research to effect change to national policy and national policy instruments for child online safeguarding in the United Kingdom, as well as contributing to national discourse and the terms of debate in this area.

a Establishing the national co-ordination of online safeguarding through the inclusion of "online safety" within statutory guidance and instruments

In 2012, Phippen's commissioned review of the SWGfL self-review data showed that both the online safety training of staff and governors in schools and colleges and recording and monitoring of online safety incidents were poor. The SWGfL fed these findings into government guidance and OFSTED reviews, which led to the inclusion of online safety within OFSTED inspector's briefing from 2012 onwards, and within the Department for Education's statutory Keeping Children Safe in Education guidance from 2016 onwards. These changes have improved the national coordination of training for those in the children's workforce, by enhancing statutory instruments (i.e. schools inspection) to ensure the implementation of national policy in practice, the roll out of which has been supported by Phippen's contribution to the materials for OFSTED inspectors in online safeguarding; 3 of the 13 questions of the OFSTED Inspecting e-Safety: Briefing for Section 5 Inspection, used by all inspectors throughout the impact period, are directly attributed to Phippen's report, published by SWGfL. Together, the briefing and the 2016 statutory guidance make online training for staff, governing bodies and proprietors a statutory and regulated requirement for all schools and colleges educating those under the age of 18 years.

We have discussed the routes to policy change earlier, and the narrative here describes its significance. While online safety seems an assumed part of statutory safeguarding and inspection in schools now, just over 10 years ago it was not the case (even though online harms research was well established in the early 21st century). Again, this would have been far more difficult to achieve if it was not for working in partnership with organisations who had direct links to those able to change policy, and the research underpinning it made it very difficult to disagree with the changes being called for.

b Shaped discourse, debate, and legislation on how to safeguard children online

Phippen is actively engaged in national discourse about keeping children safe online, providing evidence for legislation and lobbying by NGOs, helping to shape national policy debate and understanding away from reactive criminalisation of children towards effective methods of safeguarding. In the impact period, Phippen has provided written evidence directly to government including the Ministry of Justice; the Department of Digital, Culture, Media and Sport, who cite him in their Digital Economy Bill Briefing Paper 2016; POST (who cite Phippen in POSTNOTE Number 608, July 2019), as well as to statutory bodies (including OFSTED and the BBFC). Phippen has also provided NGOs with research-based evidence for lobbying government (including the UK Safer Internet Centre, the Open Rights Group, the Marie Collins Foundation, and Netsafe New Zealand). Recently co-produced reports include Young People, Internet Use and Wellbeing with the SWGfL, and Young People and Sexting – Attitudes and Behaviours – a cross-country comparison around young people's attitudes to teen sexting with SWGfL, the Office of the Australian eSafety Commissioner, and Netsafe New Zealand. Phippen is also repeatedly invited to present his research at national industry and policy events, for example, speaking at Interpol's 2017 Crimes Against Children conference, sharing knowledge on teen sexting and youth-produced sexual images to law enforcement professionals in over 30 countries.

The aforementioned reach is all underpinned by the evidence base developed in partnership with the NGOs we have discussed previously. The international work in Australia and New Zealand came directly as a result of working with SWGfL and their international partners.

Phippen also contributes to national public discourse, frequently being quoted in the UK national media, including on age verification, the use of technology by children, and the MOMO challenge (e.g. *the Guardian* (30/12/19), *the Times* (13/1/18), *the Daily Mail* (4/2/16), *the Sun* (2/2/15), *the I* (6/12/19), and *Forbes* (30/10/19)). This coverage is significant as a catalyst for change, primarily through influencing stakeholders such as parents and teachers in thinking about how they approach conversations with young people and how to reflect upon their own attitudes towards child online behaviours.

While media coverage is not something I actively court, it can be useful if one is trying to influence public opinion and the policymakers who are swayed by it. While I have always been careful to not sensationalise the work we do in order to be more likely to reach a media outlet, there have, on a number of occasions, been opportunities to engage in more serious discussion in the press, and this has certainly had some value in starting conversations with policymakers. It is certainly one of the weaker forms of engagement that can result in change, it can, at least, be useful in starting discussions.

Enhancing Legislation and Support for Victims of "Revenge Porn"

Phippen's research has directly influenced the enhancement of legal protections and support services available to victims of "revenge porn". Revenge porn is the colloquial term often used to refer to the non-consensual sharing of intimate images and other media. Phippen worked with campaigner Laura Higgins to provide evidence to successfully lobby UK Parliament for the inclusion of so-called revenge porn as a sexual offence in section 33 of the Criminal Justice and Courts Act (2015). Higgins said "[Phippen's] pioneering work on sexting provided an evidence base through which we were able to show that image based abuse was real and problematic, and created the basis for both the campaign that led to a change in legislation".

Higgins worked with MP Maria Millar to bring about this change, and was twice mentioned as a leader in this area in Parliamentary debate on the matter (Hansard, 19 and 21 June 2014). Alongside and in conjunction with the new legislation, Phippen worked with Higgins on the evidence to underpin the establishment of the Revenge Porn Helpline in 2015 – both the case for setting it up, and to win funding for it from the home office. Phippen's research provided

> the evidence to obtain funding and support to set up a helpline for those affected by this crime... It helped us understand that it was not being taken seriously as a form of abuse, and reflects the revictimisation that occurs through judgement both on social media channels and also with some stakeholders... The work Andy has done on image-based abuse from 2009 onwards allowed us to make a case that this is a modern day phenomenon that affects many people's lives and there is much judgement and victim blaming, strongly making a case for the need to support victims.

From 2015 to the end of December 2020, the Revenge Porn Helpline helped over 8500 people who have contacted it seeking support and advice for cases of intimate image abuse, and had removed over 92% of the content it reported on behalf of client.

Finally in the details of the impact, this was a very specific issue but shows how engagement, and supporting those with the ear of politicians, can have lasting, long-term impact that can change outcomes for a whole population. Until 2015, there was no specific law to prosecute someone who chose to share intimate images of someone (often an ex-partner), and while there were various pieces of legislation that could have been used (and in some cases civil law was used (Phippen and Brennan 2020)) their application was inconsistent and outcomes for victims could be extremely unsatisfactory. SWGfL, with whom I had worked on a lot of early research around teen sexting, was in an ideal position to lobby for a change in the law in order that victims might be better supported, and also call for funding for the Revenge Porn Helpline, which continues to grow to this day, in order to provide individualised support for adult victims of image-based abuse. The lobbying SWGfL conducted specifically with MP Maria Miller resulted in a change in the law that has been applied many times since its introduction in 2015. As the aforementioned testimonial highlights, a lot of the discussions with Mrs Miller and her team were informed by this early work in order to show the prevalence of practice, the lack of support available, and the impact of educational messages that resulted in victim blaming and, in some cases, an unwillingness to ask for help.

As with many of the discussions through this chapter, it would not be impossible for these outcomes to be achieved as a lone academic making the case, or even pursuing a research grant and then contacting potential stakeholders to be partners in the research. Instead, working in partnership means that such changes are far more likely to happen and that these impacts are more likely to continue. What is certain is these outcomes would have been nowhere near as effective if a post-project or post-publication approach to impact was adopted.

In bringing this deconstruction to a close, there are a number of themes that arise. First, the research has never had impact imposed upon it, it has always been a natural and important part of my research identity, and symbiotic is a term I return to a lot. This exploration showcases a deep, mutually beneficial relationship between academic research and the practical outcomes that influence real-world value in the broad realm of online safety. This symbiosis emerges from a continuous, cyclical, and dynamic interaction between generating knowledge through research and applying that knowledge to influence policies, practices, and individual behaviours.

From the early days of engaging with this research direction (which was, of itself, somewhat unintentional, as discussed in Chapter 1), the activity is fundamentally designed to be impactful by addressing practical problems and providing solutions that are grounded in evidence and often co-created with stakeholders. This approach has ensured that academic findings do not remain confined to scholarly papers but are instead integrated into practices and policies that affect real lives. This is clearly illustrated in the online harms that have directly influenced how policies are shaped to protect young people from online risks.

Conversely, the engagement with stakeholders helps to refine and direct research objectives by maintaining an ongoing dialogue with these groups, which in turn ensures that my research agenda is continuously aligned with the actual needs and challenges faced by those at the frontline of online safety. This is not to say that the research is driven by the stakeholder's agendas, but it is always conducted while being aware of debates in the wider ecosystem, which is generally informed by deep relationships with these stakeholders. This symbiotic relationship extends to the methodology of the research and by involving stakeholders directly in the research process, such as through participatory data collection or collaborative analysis sessions, and sometimes with the co-creation of tools and approaches, the methods can be informed by a wide array of perspectives and insights which enhance the depth and breadth of the findings and help produce more robust outcomes.

Finally, the impact of the work is amplified through its recursive nature and long-term relationships. Each successful implementation of research findings leads to new questions and areas for further investigation, driving a cycle of continuous improvement and innovation in the field, adapting to new challenges and technologies in the realm of online safety or, as is increasingly the case, contradicting the moral panic around emerging technologies and a reminder that the core practices should remain the same.

Consequently, the symbiotic nature of this work not only enhances the impact of my research within academic circles but also ensures that it contributes tangibly to societal well-being, demonstrating *a* model for how academic research that can effectively intersect with and influence real-world outcomes. As we stated in opening this chapter, we are not saying this is *the* way to do impactful research, but it is an approach that has proven to be effective. And it shows that deep impact does not come from chasing the case study, it comes from caring about the research and how it can change people's lives.

Conclusions

This chapter has showcased a model of how academic research can meaningfully intersect with public policy and societal needs, illustrating the dynamic process of impact generation that has an intrinsic centring upon the identity of the academic conducting the work. In conclusion, the case illustrates the difference between conducting a series of academic exercises or obligations because management or policy dictates it, but one that was often in conflict with senior management direction yet was still maintained, due to a personal belief that it was important to integrate research with real-world application in an area that aimed to improve outcomes for young people. The accomplishments highlighted throughout the chapter underscore the value of this approach, particularly in how evidence-based practices can significantly enhance public policies and educational strategies, ultimately fostering safer online environments for children and young people, while still remaining academically sound.

The long-term collaboration with various stakeholders, including schools, government bodies, and NGOs, has not only expanded the reach of the research but also deepened its impact. This approach has enabled a reciprocal flow of knowledge and resources, ensuring that the research is continually informed by real-world challenges and opportunities. And this symbiosis has led to the development of initiatives that are both innovative and grounded in robust evidence, demonstrating the potential for academic work to drive substantive change.

It is, hopefully, an illustration that impact does not have to be driven from management or policy whim but instead forms a core part of academic identity. We would argue that without this belief in the work, or its place within academic practice, it would have been given up upon a long time ago with a refocus on whatever management direction was prevalent and most likely to elicit institutional reward. However, the integration of research and practice as demonstrated in this chapter serves as a powerful example of how academic endeavours can challenge traditional boundaries and have a profound impact on society. The lessons drawn from this work can emphasise the importance of persistence, collaboration, and a deep commitment to societal well-being, principles that can guide future efforts in academic research and its application in addressing complex social issues. This model not only enriches the academic field but also contributes tangibly to improving lives, particularly in the realms of education and public policy.

Notes

1 www.headstartkernow.org.uk/digital-resilience/ [Accessed June 2024].
2 www.swgfl.org.uk/ [Accessed June 2024].
3 https://revengepornhelpline.org.uk/ [Accessed June 2024].
4 www.farrer.co.uk/ [Accessed June 2024].
5 www.360safe.org.uk/ [Accessed June 2024].

References

Bronfenbrenner, U. (1979). *The ecology of human development: Experiments by nature and design*. Harvard University Press.

Farrer and Co. (2023). Addressing child-on-child abuse – A resource for schools and colleges. www.farrer.co.uk/globalassets/clients-and-sectors/safeguarding/addressing-child-on-child-abuse.pdf [Accessed June 2024].

Huzzard, T. (2021). Achieving impact: Exploring the challenge of stakeholder engagement. *European Journal of Work and Organizational Psychology*, *30*(3), 379–389.

Lacohée, H., Phippen, A. D., & Furnell, S. M. (2006). Risk and restitution: Assessing how users establish online trust. *Computers & Security, 25*(7), 486–493.

Nutt, D., King, L. A., Saulsbury, W., & Blakemore, C. (2007). Development of a rational scale to assess the harm of drugs of potential misuse. *The Lancet, 369*(9566), 1047–1053.

OFSTED (2023). Education Inspection Framework. www.gov.uk/government/publications/education-inspection-framework/education-inspection-framework-for-september-2023 [Accessed June 2024].

Phippen, A. (2016). *Children's online behaviour and safety: Policy and rights challenges.* Springer.

Phippen, A. (2023). England Schools Online Safety Policy & Practice Assessment 2022. https://swgfl.org.uk/assets/documents/online-safety-report-2022.pdf [Accessed June 2024].

Phippen, A., & Bond, E. (2023). *Policing teen sexting: Supporting children's rights while applying the law.* Springer Nature.

Phippen, A., & Brennan, M. (2020). *Sexting and revenge pornography: Legislative and social dimensions of a modern digital phenomenon.* Routledge.

Phippen, A., & Street, L. (2022). *Online resilience and wellbeing in young people.* Springer International Publishing.

Reed, M. S., Bryce, R., & Machen, R. (2018). Pathways to policy impact: a new approach for planning and evidencing research impact. *Evidence & Policy, 14*(3), 431–458.

UK Government. (2024). Keeping Children Safe in Education. https://assets.publishing.service.gov.uk/media/6650a1967b792ffff71a83e8/Keeping_children_safe_in_education_2024.pdf [Accessed June 2024].

3 Public Good, the Civic University, and Other Policy Ideas

Introduction

We now move from the focused, individual level, to return to the policy area and start our journey to address why research impact is not well understood, or regarded, by many in academia, and why it should be. While this exploration, we acknowledge, is a little dry compared with the explorations that take place later in the book around our empirical work, this is a necessary starting point because we cannot really argue against the policy direction if we have not, at first, explored from where it emerged and, more generally, illustrated the gulf between policy interference in academic practice and its likely impact upon both identity and the ways in which academics work.

In exploring the culture around research impact, its motivations, and barriers to being effective, it is worthwhile first to situate the concept of research impact against the changing policy landscape in higher education and where the current (as written in 2024) leading edge of university outreach, to use a broad term, lies, and whether current policy positions do anything to help us better understand first institutional understanding around research impact. And also why, in our experiences, and those of many of those we have spoken to as part of this project, feel that research impact as a part of academic identity is still misunderstood and not embedded in broader academic practice.

As such, we will spend some time in this chapter looking at policy evolution within, mainly, the UK higher education section, but also explore cultural shifts in higher education as a whole, moving away from the liberal and civic intentions first introduced at the end of the 19th century to neoliberal approaches to higher education (and arguably the marketisation, metrification, or instrumentalism of academic endeavour (Furedi 2010)) which, it has been argued (Molesworth, Scullion and Nixon 2011), beset current higher education practice. It will also explore whether modern civic concepts, which centre on evidencing the role of an institution in its local environment, and other frameworks, such as the United Nation's Sustainable Development Goals, have helped or hampered the embedding of research impact into academic

DOI: 10.4324/9781003454991-3

practice and start to explore whether these top-down directives result in any improvement of core academic practice.

Furthermore, we will spend some time within this chapter introducing the notion that unless research impact forms part of core academic identity, frameworks and management imposition will do little to enhance how research findings might impact upon wider society (which surely is the goal of any impact agenda?).

Making Capable and Cultivated Human Beings

The quote given next comes from John Stuart Mill's inaugural address (Mill 1867) delivered on his installation as Rector of the University of St Andrews in 1867:

> The proper function of a university in national education is tolerably well understood. At least there is a tolerably general agreement about what a university is not. It is not a place of professional education. Universities are not intended to teach the knowledge required to fit men for some special mode of gaining their livelihood. Their object is not to make skilful lawyers, or physicians, or engineers, but capable and cultivated human beings.

This quote is frequently used by many who write about higher education policy and public good (e.g. Williams 2016; Donner, Urbinati and Zakaras 2007; Ward 2023) as an illustration of how modern neoliberal policy imposing metric-driven assessments of targets such as employability, contact hours, student survey scores, and research metrics has lost sight of the overall goal of intuitions, which is to develop citizens who are sufficiently knowledgeable and informed to make a positive contribution to a liberal society.

It is argued (Barnett 1994) that there was once a clear understanding of the role of the university in society, and John Stuart Mill's speech at St Andrews is frequently used as a lynchpin of this ideology (Willaims 2016). While perhaps there is some rose-tinted historical gazing with this view, it is unquestionable that the speech is a significant historical document that had a profound influence on the perception of the role of universities.

Mill, a renowned philosopher and economist, from a classic liberal tradition, used this platform to articulate his vision of the purpose and responsibilities of higher education institutions that perhaps resonates with academic identity in a manner that measurement, management imposition, and prescriptive policy do not. Certainly, his ideas have had lasting impacts on educational philosophy and the concept of the public good of universities.

It laid the groundwork for a vision of higher education that goes beyond the mere transmission of knowledge to the next generation of workers to encompass a more holistic development of individuals and their ability to contribute to society as a whole. While he certainly acknowledged the importance of producing professionals, his view was more that through education, critical thinking, and diversity of opinion citizens would be better equipped to become socially aware and progressive thinkers within their professions, rather than simply be upskilled from undergraduate to someone who would slot into a given role. He articulated the moral and civic responsibilities of universities that continue to influence contemporary discussions on the role of higher education in society and its contribution to the public good.

The concept of the public good in relation to higher education remains one of debate to this day (e.g. Cantor 2020; Olssen 2021; Busch 2023), but centres on the concept that universities and other educational institutions contribute to the welfare and improvement of society as a whole, rather than just serving the private interests of individual students. While a student might attend university in order to become better qualified, in order to develop better career prospects, it is argued that higher education has to serve a greater purpose than offering upskilling programmes for individuals. There is a view, as articulated by Mill in his address, that while students might be accessing education related to a profession, the learning they receive should more broadly help them become more critical thinkers and understand the society in which they will live and work.

Mill emphasised the importance of higher education as a means of intellectual development. He argued that universities should not merely teach professional skills but should aim to cultivate a robust and versatile intellect. This development was seen as crucial for enabling individuals to understand, engage effectively with and contribute to society.

He also highlighted the broader societal benefits of higher education, suggesting that educated individuals are better equipped to perform their duties as citizens (their civic role). This, he believed, was vital for the health and progress of a democratic society. By fostering critical thinking and moral integrity, higher education should produce citizens who are thoughtful, informed, and capable of making decisions that benefit the public interest rather than simple personal benefit.

He further stressed the idea that universities should promote not just individual success but also the pursuit of truth and the general welfare. He saw the university as a place where free and open discussion should be encouraged, and where ideas should be exchanged without prejudice. This environment would contribute to the development of more enlightened and morally grounded citizens who would be able to debate and understand different viewpoints from members of society, rather than being fixed in their views and their politics.

These deep civic concepts informed late 19th century thinking about the civic role of higher education. A graduate environment enriched with deep civic concepts offers a more stable and enduring foundation for individual development and for the betterment of society. Critical thinking skills empower individuals to analyse information, evaluate arguments, and solve complex problems beyond the confines of any single discipline or current market trend. Social understanding fosters empathy, collaboration, and the ability to navigate the diverse social landscapes of an increasingly globalised world. And liberal philosophy, with its emphasis on ethics, justice, and the exploration of human values, equips students to question, innovate, and lead in a way that mere technical skills cannot.

It has been argued (Okolie et al. 2020; Dekker 2020) that education which prioritises these foundational competencies over transient skills do not merely prepare students for their first job but for a lifetime of careers, many of which may not yet exist, and are essential in modern society, in that it cultivates adaptability, resilience, and the capacity for lifelong learning, attributes that are increasingly recognised as critical in the face of rapid economic and technological change. However, in more recent policy debate around the role of higher education, there has been a far greater emphasis on employability (Tomlinson 2017) that has now raged for as long as we have worked in the sector. There is an, arguably, neoliberal view that seems to not have specific political colours, that if public money is spent on something (e.g. a loan for a student wishing to undertake a degree), then society should see a return on investment, such as a more effective contribution to the workforce. However, these views miss the importance of liberal and critical education as essential life skills in developing a society, far beyond knowing a specific programming language or an alignment with the demands of a professional body. Of course, we are not saying there is no role for employability in degrees, but this should not be the only goal of higher education and therefore public good and civic responsibility are as crucial now (some might say even more so) than when Mill first delivered his address.

There are a number of ways in which higher education can contribute to the wider society in which institutions sit:

- **Economic benefits:** Higher education contributes to economic development by developing a skilled workforce and a pipeline for the future workforce, while fostering innovation through research and development, and increasing economic mobility for individuals. The higher earning potential and increased productivity of graduates will certainly provide individual benefits but can also lead to broader economic growth and stability.
- **Civic engagement:** Educational institutions often play a crucial role in fostering civic engagement and responsibility. Graduates are generally more likely to vote, participate in community service, and engage in public discourse, contributing to a healthier democracy.
- **Social benefits:** Higher education can promote social mobility and help reduce social inequalities by providing opportunities for individuals from diverse backgrounds to advance and succeed through widening participation policy and practice.
- **Cultural development:** Universities can be seen as centres for cultural preservation and innovation, offering spaces where artistic, philosophical, and scientific pursuits can be explored and flourish.
- **Research and innovation:** Higher education institutions are pivotal in advancing knowledge and innovation. Research conducted within

universities can lead to breakthroughs that solve societal challenges, such as disease, environmental issues, and technological needs, and these findings can be shared, through education, with students who will take this new and emerging knowledge into wider society and workplace practice.
- **Ethical development:** Higher education that includes a focus on ethical reasoning and moral development can help graduates understand the broader impacts of the actions of citizens, governments, and businesses on society which can lead to a more ethically aware populace.

However, there are views (Lynch 2006) that suggest that the neoliberalisation of higher education (discussed in far more detail later) has resulted in us losing sight of Mills vision in exchange for a marketised product offer in a competitive and financially volatile modern higher education environment (see further), rather than understanding the need for deep civic concepts (Giroux 2010).

Deep civic concepts refer to ideas and principles that are fundamental to understanding and engaging with civic life, governance, community involvement, and the responsibilities of citizens within a society (Ostrander 2004). These concepts go beyond basic knowledge of governmental structures or civic processes; they delve into the philosophical, ethical, and practical underpinnings of what it means to be an active, contributing member of a community or society. Deep civic concepts are, some argue (Checkoway 2001; Barnett 2016), crucial for fostering a sense of civic responsibility, community engagement, and the development of a healthy, democratic society. Yet it is arguable whether a modern higher education sector still embraces these values.

Nevertheless, this book, in considering the role of research in academic identity, and specifically the need to engage research findings beyond the academy, considering whether research impact is a fundamental aspect of this, or whether it is an imposed aspect of the modern university environment is vital. This sits alongside the question of whether there might be a way which it could be part of academic identity without it being viewed as an imposition. As such, it is also worth reflecting on how Mill's address, with its focus on the public good of universities, relates to research within higher education.

The address did not explicitly focus on the topic of research in the modern sense of the term, particularly as it relates to scientific inquiry, the production of knowledge within academic disciplines, and its dissemination through academic journals. However, his speech did touch upon themes relevant to the pursuit of knowledge and intellectual inquiry, which are foundational to the concept of research in higher education.

Mill emphasised the importance of a broad and liberal education, arguing that the development of the mind and character was paramount. He believed in fostering intellectual independence and critical thinking, encouraging students to engage deeply with a wide range of subjects. This approach to education implicitly supports the cultivation of a scholarly mind-set that is curious, questioning, and open to new ideas – qualities that are essential for research and ones that, most would agree, are essential aspects of the academic identity, which might highlight why there has been some tension with an imposed impact agenda (Smith et al. 2020) from policymakers, research funders, and senior management.

Moreover, Mill's advocacy for freedom of thought and expression within educational institutions can be seen as a precursor to the modern emphasis on academic freedom, and debates about its erosion through increased management interference and research assessment (Curtis 2011), which is crucial for a thriving research environment. The ability to explore, question, and challenge existing knowledge without fear of censure is fundamental to the advancement of scholarship and innovation.

While Mill's speech might not have directly addressed the mechanics of research or its role within the university, his principles about education, intellectual development, and the free exchange of ideas are fundamentally supportive of the research mission of contemporary higher education institutions. These principles underline the importance of creating an academic culture that values and promotes rigorous inquiry, scholarly excellence, and the generation of new knowledge.

In the contemporary discourse on higher education, the emphasis on employability and the alignment of educational outcomes with the immediate skills needs of the economy has become increasingly prevalent. While the importance of preparing graduates for the job market is undoubtedly a crucial aspect of teaching and learning within higher education, a myopic focus on employability, understood merely as the acquisition of market-driven skills, risks undermining the broader, enduring value of higher education and fails to grasp Mill's appreciation of how a liberal critical education is as important for graduate transition that being skilled in a particular profession. While there have been some arguments reviewed that academics cannot be publicly funded to perform research of personal interest with no social or economic value (Cooksey 2006; Radder 2010), we would argue that is a clear relationship between academic research and the public good without the policy interference (Pusser 2006; Marginson 2011; Edwards and Roy 2017).

Advancing knowledge for societal benefit: It is essential that universities remain knowledge-creating institutions, else that may as well rebrand as training centres or further education colleges. The research they conduct often leads to new discoveries and innovations that can have broad benefits. There is public good in studies that can, for example, lead to new medical treatments, inform public health guidelines, contribute to environmental sustainability, or enhance the effectiveness of educational practices.

Informing policy and public debate: Research from universities often plays a vital role in shaping public policy and discourse, currently this is perhaps best seen through both climate change (Dilling and Lemos 2011) and the COVID pandemic, with its reliance on academic research to inform policy direction (to an arguable degree Stevens 2020). When research findings are translated into policy recommendations, they can have significant societal impacts, such as improving social welfare, shaping economic policy, or influencing legal reforms. The public good is served when research informs legislation or public understanding of critical issues, leading to informed decision-making and improved societal outcomes.

Economic Development: Universities contribute to the public good through research that drives economic development. This can include the creation of new technologies, the establishment of start-ups, and contributions to local and national economies through job creation. Universities that foster relationships with industry and government can help to translate research into products, services, and practices that contribute to economic growth.

Cultural enrichment and addressing societal challenges: Humanities and arts research can enhance cultural understanding, preservation, and expression, and we explore some cultural tensions with the understanding of the role of the humanities in Chapter 5. Research can also help address major societal issues like inequality, poverty, and injustice, providing a foundation for interventions and programmes that improve the quality of life for marginalised groups.

Education and training: Part of the public good is creating an educated populace. University research contributes to this by developing new pedagogical techniques, contributing to curriculum design, and providing evidence-based approaches to learning and teaching. Furthermore, it should be an expectation within a university that the education it delivers is informed by research findings to make it as contemporary as possible. This not only improves the educational outcomes of students but also ensures that graduates are well-equipped to contribute to society.

If we specifically consider research impact, given that is the focus for this book, we can see that the majority of these issues can be directly related to research that impacts upon the stakeholder community, given that it encompasses the ways in which research contributes to innovations, supports policy-making, and generates societal improvements. For example, the development of new technologies and products, the formation of strategies to address social and environmental issues, the development of new, evidence-based, legislation, and the enhancement of public services and systems should all be considered facets of research impact. Research outcomes can directly influence public health policies, leading to better healthcare provisions and disease-prevention measures. Economically, research can contribute to growth by fostering new industries and improving existing ones through technological advances. And in the environmental sector, research findings help in crafting sustainable practices and policies, mitigating the impacts of climate change and preserving natural resources. Ultimately, research impact should be viewed as the manifestation of how scholarly activities contribute to the broader goals of societal development and well-being. And how it serves the public good. Which does beg the question regarding why, if this is the case, the concept of research impact remains poorly understood in institutions and resisted by a lot of the academic community.

Civic Universities, KEFs, and SDGs

In answering this question, we feel it is worthwhile to review the common model for what we might refer to as civic engagement in higher education, which tends to be top-down in nature and driven not by motivations of public good but instead value for money. In this section, we explore some emerging policy positions that have an influence on the way universities in the United Kingdom are expected to work with external stakeholders, how these processes are expected to be carried out, and how they are measured.
Given the UKRI definition of research impact[1] is:

> an effect on, change or benefit to the economy, society, culture, public policy or services, health, the environment or quality of life, beyond academia.

One might expect all of these external drivers to contribute to the adoption of research impact as part of academic practice, given they all ask universities to look outwards and demonstrate their value either to their local, national, or global communities. However, the nature in which these expectations are articulated might be in tension with academic identity and motivation, as we will explore in more detail later in this book.

The concept of civics related to higher education is, some would argue, enjoying something of a renaissance in the university sector, with the establishment in the United Kingdom of the Civic University Commission[2] and the publication in 2019 of the Truly Civic Report (UPP Foundation 2019), which we will explore in more detail later.

However, as discussed earlier, the civic concept was explored in detail in Mill's address in the 19th century, and the concept of a "civic university" was first developed in the United Kingdom during the late 19th and early 20th centuries. These institutions, sometimes referred to as "red brick" universities, were distinguished from older, more traditional universities like Oxford and Cambridge. The term "civic university" was associated with universities that were founded or received significant development in major industrial cities and were intended to provide higher education with a strong emphasis on meeting the needs of their local communities, particularly in technical and scientific fields – similar to the civic mission of some universities in the present day.

The first civic universities included institutions like the University of Manchester, University of Birmingham, University of Liverpool, and University of Leeds. These universities typically were based in industrial centres and had close ties with local industries, focusing on applied science and engineering disciplines to support and enhance local economic development. The civic university model emphasised accessibility, relevance, and service to society, aiming to extend educational opportunities beyond the traditional elite classes and contribute directly to the social and economic vibrancy of their regions.

However, the modern civics approaches, while still relating to these underlying concepts, seem to have emerged not from a wish for further public good, but value for the public purse.

In the United Kingdom, the Civic University Commission was established to explore how universities could more effectively engage with their local communities and serve the needs of the society around them. The commission was launched by the UPP Foundation in 2018 and aimed to help universities realise their potential as civic institutions, thereby contributing positively to the places where they are located.[3]

The key focus of the commission was to understand and enhance the role of universities in their cities and regions, addressing social and economic challenges through education, research, and public engagement. The commission gathered evidence, engaged with stakeholders, and produced reports with recommendations on how universities could strengthen their civic roles. It emphasised the importance of universities acting not just as educational bodies but as integral parts of their communities, contributing to local development and policy-making in meaningful ways.

The Civic University Commission was chaired by Lord Bob Kerslake, a prominent figure in British public life and former head of the civil service, and an advocate for the civic role of universities.

Alongside Lord Kerslake, the commission included various representatives from the higher education sector, local government, and businesses, ensuring a wide range of perspectives were considered. Notably, the commission involved several university vice-chancellors and other educational leaders who provided insights into the practical aspects of enhancing universities' civic engagement. However, it is perhaps telling that academic representation on the commission tended to be at a senior management level, rather than a representation from those active in the academy who might, we would argue, bring a more authentic academic representation to the discussions.

The commission also collaborated with key educational organisations and stakeholders to enrich its understanding and recommendations. This collaborative approach allowed the commission to draw on a broad base of expertise and experience, ensuring its findings and recommendations were both robust and relevant to diverse communities across the United Kingdom.

The final report of the commission, the Truly Civic Report, aimed to contrast between the civically engaged university and the truly civic university, and stated that (UPP Foundation 2019):

> A true civic university has a clear strategy, rooted in analysis, which explains what, why and how its activity adds up to a civic role. Whether it does that through leveraging international activity or focusing locally; primarily as a research and teaching institution or through a wider anchor role; and

alone or in an ecosystem is a local question. But it should be clear why and how universities have answered that question, and how they have organised themselves to achieve their civic aims.

And key recommendations from the report included

- **Civic university agreement:** Encouraging universities to formalise their commitment to their civic roles by defining a policy document to direct their civics strategy.
- **Measuring and incentivising success:** Developing metrics to evaluate the impact of universities' civic activities.
- **Funding the civic:** Proposing the creation of a Civic University Fund and advocating for increased funding through existing funding schemes.
- **Spreading good civic practice:** Sharing effective strategies and practices among universities.
- **Civic widening participation and attainment:** Focusing on local engagement and tailored community interactions.
- **A resurgence of adult education:** Highlighting the importance of lifelong learning, particularly in response to changes such as automation and artificial intelligence.
- **Strengthening local impact alongside international excellence:** Balancing local contributions with global academic standing.
- **Role models as employers, procurers, and brokers:** Universities acting as leaders in their economic ecosystems.
- **Catalysts for local economic growth:** Leveraging university resources to stimulate local economies.
- **A broadened strategy on culture:** Using cultural activities to enhance community engagement and development.

The report also led to the establishment of the Civic University Network, which supports universities in achieving these goals by providing resources, sharing best practices, and fostering collaboration among institutions aiming to enhance their civic impact.

Within the report there were also a number of tests defined to determine whether a university might be considered civic:

A public test

- Can people talk about "our university" with pride and awareness?
- Is civic activity aligned to public wants?
- Are the views of local people reflected in either the formal governance or informal and communication structures and strategies of the university, including as regards the progress against the goals of the Civic University Agreement?

A place test

- How well are the university's teaching programmes aligned with the structure and demands of the local labour market and likely developments in the coming years?
- Which population is the university serving? How local, and how diverse (including in terms of age, gender, race, ethnicity, sexual orientation)?
- Are universities following a single national university model or are local variants being adopted?

A strategic test

- Can the university define where its civic boundaries are? What is the geographical area it is primarily focused on?
- Is it clear how the university is linked to local/regional leadership in the area – including its graduates?
- Does the university have a clear analysis and strategy? Can it use that analysis to explain/underpin why its global or national activities supports/strengthen its civic activity, and can it clearly articulate what outcomes, impacts, and benefits it wants as a result of its strategy?

An impact test

- Can the university measure the impact of its activity?
- Has the university been thoughtful about how it works with other local universities and other institutions to maximise that impact?

While we can see there are some parallels with civic goals, as set out in the Truly Civic Report, and arguably there should indeed be parallels between policy that aims to embed knowledge from an institution in its community, there equally seem to be few crossovers between research assessment policy (where, as we have explored in Chapter 1, a lot of the drive for research impact has arisen) and civics. However, we might also observe that the goals of measurement, agreements, incentivisation, and policy levers (e.g. associated funding) do have some similarities with the manner in which research impact has evolved from assessment policy, which is less about the belief in the importance of impact, or understanding the motivations for adopting such practice, and, instead, imposing policy and management structures as the motivation to adopt such practices. Furthermore, there are challenges of civics being placed in the local space which can be viewed (including in the Truly Civic Report) as second-class research compared with that which has global impact. While reach and significance, core "metrics" in research impact which we will discuss in Chapter 4, are hotly debated there is a view in our experience and that of those we interviewed (see chapters 5 and 6) among academics and research manager that global research is "better" than local remains.

Furthermore, this civic university approach once again proposes a prescriptive model driven by policy and "tests" to see if you tick sufficient boxes to be a civic university. While this is, of course, a reasonable method at looking at civic engagement at a strategic level, it does little to embed these concepts across the institutional workforce.

Certainly, the core concepts of civic practice in higher education *should* have a relationship with research impact and the public good. However, we would question whether current policies reflect this deep civic concept or has the civic university concept lost its way? We would argue that an approach that focusses first on policy and measurement, rather than motivation and support, will only ever be shallow in its civic approaches.

There is a risk, in promoting the civic concept from a position of process and prescription, that civic responsibilities have become university administrative function. For example, in our experience and in discussions with peers in the production of this book that the production of civic university agreements which seem to be almost completely disconnected from faculties and schools, much like Corporate Social Responsibility in the private sector. We would foresee that the impact (no pun intended) of this policy work on universities will not be transformative. While the focus of most of our Freedom of Information work conducted for this book lay in practices to support research impact in institutions, we did also ask whether institutions had a civic university agreement (one of the key recommendations from the report). In total, out of 109 institutions, 31 said they did.

Where there is crossover with the academic community, it is once again as a result of evaluation frameworks, such as the rather bewildering Knowledge Exchange Framework (KEF),[4] which universities seem duty bound to carry out with little enthusiasm or engagement, often in isolation from the academic community that will be carrying out a lot of the sort of work that should be returned.

The KEF is a recent addition to the pantheon of assessment frameworks for universities to return to government and funders. It is stated by Research England[5] that the KEF aims to enhance how English universities engage in knowledge exchange activities, promoting efficiency and effectiveness in using public funding. Furthermore, it aims to encourage a culture of continuous improvement within higher education institutions by providing detailed, publicly accessible data. KEF allows universities to assess and enhance their performance in knowledge exchange, therefore fostering a deeper understanding of their contributions to the economy and society.

The KEF evaluates universities based on several key perspectives including research partnerships, collaboration with businesses, work with the public and third sector, and intellectual property commercialisation. This framework is designed to compare universities in a non-competitive manner, grouping them into clusters of similar institutions to ensure fair and relevant comparisons and motivate better practice in knowledge exchange which will, in turn, improve value for money for the public purse.

Similar to civic agreements, it has been proposed that KEF seeks to transparently showcase the diverse ways universities interact with external partners and stakeholders, aiming to bolster their roles in local growth and societal advancement.

However, with no core funding associated with it (in contrast to quality-related (QR) funding that is an outcome of the Research Evaluation Framework), there seems to be little motivation for university senior managers to engage with the process and it is, instead, passed to middle management professionals who will compile and send off the return. Certainly, in our own experiences, and in conversations with peers interviewed for this book, there seems to be little link between academic practice and the KEF. This is the case in our own experiences, and that of those academics who we interviewed regarding their research impact. All of these academics had significant reach across their stakeholder spaces and engaged with industry, civil society, and communities. Yet none of those we interviewed had been involved in their institutions' KEF strategy.

Returning to the global context, this is also not without its policy influences, key to these being the United Nations Social Development Goals, which seem, at present, to be rife in all facets of academic policy, if not practice.

The United Nations Sustainable Development Goals (SDGs)[6] are "a universal call to action to end poverty, protect the planet, and ensure that all people enjoy peace and prosperity by 2030". These goals were adopted by all United Nations Member States in 2015 as part of the 2030 Agenda for Sustainable Development, which set out a 15-year plan to achieve the goals. It has been stated that universities have a critical role to play in achieving these goals,[7] given their influence and resources in education, research, community engagement, and policy advocacy. UK Research and Innovation (UKRI), as a central body that brings together the UK's seven research councils, Innovate UK, and Research England, often includes specific calls that are designed to address global challenges in areas such as environmental science, health, and development.[8]

As with issues emerging from policy directions underpinned by civics, we would also observe that there is nothing, per se, that is problematic about aligning some research funding to global sustainability goals and, it could be argued, should as a result of this alignment, result in strong research impact. However, a fundamental concern remains that if research funding becomes too strongly aligned with SDGs, will this result in restrictions on the academic freedoms to explore what one is curious about, through one's own research track record and inclination within a research discipline? Might this closer alignment with SDGs result in academics becoming encouraged (or instructed) to only bid for research funding with the alignment, closing off more theoretical or even whimsical areas of research that might ultimately have a long-term impact upon society as a whole. There is potential that "SDG chasing" will result in academics adopting a "project"-based approach to research, where they will bid for work based upon these SDGs, rather than their own research discipline or critical mass. The problematic nature of project research is returned to later in this book.

We use the examples of civic universities, the KEF, and UN SDG not as definitive drivers for external engagement in the higher education sector, but to illustrate the point that rather than the motivation for engaging with stakeholders coming from academic will, but from external pressures, one might argue, have been imposed upon those who are then expected to perform these duties. They also, generally, have very much a metric and measurement approach to external engagement – "how do you know you are doing it properly?" or "do our test to show how civic you are". Again, this approach seems to restrict the creativity and personal motivation that should underpin an academic's impetus to communicate the research passion to external stakeholders. Being told to do something because management have been told by government to do this seems to be very much add odds with the academic freedoms that have underpinned the higher education sector for hundreds of years.

The Spectre of Neoliberalism

From the outset of this text, our core question is why research impact is not better understood by institutions, and why academics, in general, do not see it as part of their identity.

In this policy exploration we have explored some of the broad ideological positions around the role of universities in society, and suggested that one of the challenges in getting academics to understand how impact can be part of their identity is the view that the impact agenda is one that has been imposed upon them by funders and policymakers and, in turn, managers and senior leaders (once these senior leaders could see the financial motivation for adopting impact as part of institutional practice). It is our view, and one that was shared with many of those we interviewed, that unless impact is part of your academic identity, these top-down levers are not good motivators for "becoming impactful". Moreover, having a top-down imposition (and, as some told us, one that does not come from a position on knowledge) creates opposition and a dismissing of impact not as a fundamental aspect of the public good but as something you just have to do, because we told you to. For those we spoke to who we might refer to as "impact pioneers", who were doing impactful research long before it was thought about by funders or policymakers, they were doing it because *they* thought it was important, and in several instances this was in opposition to the direction they were being told to follow by leaders (who would generally, at that stage, place far greater value on outputs and income as the only measures of research that mattered). For these pioneers, the motivation to do impact was not top-down, it was internal, and a core part of their academic identity. They did it because they knew that the findings of their research need to reach stakeholders if they were going to change things. They were not doing impact because they had to, they were doing it because they wanted to.

So if this was the case for *some* academics, why are there still so many who do not see impact as part of their research, why is there a great deal of opposition from academics to doing impact, and why is it still so poorly understood by institutions?

We have, during this discussion, touched a few times on what we might refer to as the spectre of neoliberalism in higher education and there have been suggestions by some authors (Olssen and Peters 2005) that neoliberal practices in the sector have stifled intellectual curiosity and hamstrung what it fundamentally means to be an academic (Thornton 2009).

Broadly, neoliberalism is a political and economic philosophy that emerged in the late 20th century, advocating for free-market capitalism, minimal government intervention in the economy, deregulation, privatization of state-owned enterprises, and reduced public expenditure on social services. It is based on the belief that free markets and individual entrepreneurship are the primary drivers of economic growth and personal freedom.

Influenced by the work of economists like Friedman and Hayek (Jones 2014), neoliberal thought advocates for policies that reduce the role of the state in economic activities. The implementation of neoliberal policies has been particularly notable in countries like the United States and the United Kingdom during the administrations of Ronald Reagan and Margaret Thatcher, respectively.

However, the many critics of neoliberalism argue that while it may lead to growth and efficiency, it can also increase inequality and reduce the state's ability to provide services to its most vulnerable citizens.

In higher education there has been comment about how modern management and metric-driven higher education policy and leadership (Harris 2005) has a strong neoliberalism foundation. Authors (Giroux 2002) state that neoliberalism has significantly impacted higher education across various dimensions including increased marketisation and competition which leads to a greater focus on efficiency, performance metrics, and consumer-oriented approaches to education. Furthermore, there has been significant concern about policy level scrutiny and political cycles resulting in many reviews of the sector commissioned by governments (Foskett 2010).

While the recent Civic University concepts are a clear illustration of this national policy steer to influence academic autonomy, this has certainly not been the only national policy initiative focussed upon the higher education sector in the United Kingdom. Governments tend to review higher education policy. And they do it a lot. By way of illustration of this constant policy attention, in the past two decades, the landscape of UK higher education has been shaped and reshaped by several significant policy reports and reviews:

- **The Dearing Report (1997)[9]:** Though slightly older than 20 years, the National Committee of Inquiry into Higher Education, known as the Dearing Report, has had a long-lasting impact on the landscape of UK higher education, recommending tuition fees for students, which were introduced in 1998.

- **The Lambert Review of Business–University Collaboration (2003)[10]:** This report aimed at enhancing collaboration between businesses and universities in the United Kingdom, emphasising the importance of this relationship for innovation and economic growth.
- **The Leitch Review of Skills (2006)[11]:** While not solely focused on higher education, this review emphasised the critical role of higher education institutions in developing the UK's skill base and advocated for increased participation in higher education and training to meet the UK's skill needs by 2020.
- **The Browne Review (2010)[12]:** Officially titled "Securing a Sustainable Future for Higher Education", led to significant changes in tuition fees and funding models for universities in England. The report recommended removing the cap on university fees and replacing direct government funding for tuition with loans that students would pay back after graduating.
- **The Finch Report (2012)[13]:** Focused on expanding access to published research findings, the Finch Report led to a push towards open access publishing within the United Kingdom, influencing how research outputs are disseminated and accessed.
- **The Nurse Review (2015)[14]:** Conducted by Sir Paul Nurse, this review focused on research councils in the United Kingdom. It recommended reforms to ensure that the UK's research system could effectively respond to significant challenges, leading to the creation of UK Research and Innovation (UKRI) in 2018.
- **The Augar Review (2019)[15]:** Officially titled "Review of Post-18 Education and Funding", this comprehensive report recommended changes to the funding of post-18 education, aiming to make the system fairer and more sustainable. Among its recommendations were a reduction in tuition fees and the extension of the loan repayment period.

While not all of these policy reviews relate to research (and by implication research impact), they are good evidence of the scale of government interest and "interference" with higher education practice over the last 20 years, and the increased marketisation of the sector. It is also a useful illustration of the rise of neoliberalism in higher education and, if nothing else, highlights the almost constant tampering of higher education policy in the United Kingdom by governments. Since the turn of the century, there seems to be a new review approximately every three years and, perhaps more interestingly, as people who have worked in the sector over this whole time (aside from "strategy initiatives" that go largely ignored by members of the academic community in an institution), we have seen little change to the core business of universities in which we have been employed, and those with whom we have worked – teaching and research are still carried out by core staff, while policy initiatives and funding pockets are supported by research/innovation departments whose activities remain often ignored by academics.

This shift towards neoliberal influences and the number of external actors attempting to influence academic practice have prompted debates among academics about the implications for academic freedom, the quality of education, and the role of universities in society (Giroux 2002). These impacts of neoliberalism on higher education reflect broader societal shifts towards market-driven principles and away from viewing education as a public good – very much at odds with Mill's often-cited address about the core values of higher education that introduced this chapter. The consequences of these changes are complex and have sparked significant debate regarding the purpose, accessibility, and quality of higher education.

Brown (2019) has critiqued neoliberalism's influence on higher education by highlighting its transformation of education into a market-driven entity. She argues that this shift commodifies education, prioritises economic goals over democratic or intellectual ones, and undermines the essence of higher education as a public good. She has discussed how neoliberalism erodes the values of critical thinking and civic responsibility and reshapes universities to function more like businesses than spaces for learning and critical inquiry.

Similarly, Giroux (ibid.) argues that neoliberalism's influence on higher education erodes its democratic foundations, turning education into a commodity rather than a public good. He argues, similar to Brown, that this shift undermines the role of education in fostering critical thinking, civic engagement, and ethical reasoning, making it instead serve the interests of capital and corporate power. Giroux emphasises the need to resist neoliberal policies in education to preserve its role as a crucial public sphere for developing critical citizenship and democratic life.

And Watermeyer and Tomlinson (2018), whose work was so pivotal in the arguments we presented in Chapter 1, have articulated a critical perspective on the impact of neoliberalism in higher education, particularly focusing on how neoliberal policies influence academic life and societal engagement within universities. Their work is concerned with how neoliberalism has transformed higher education into a competitive marketplace, emphasising performance, efficiency, and market-driven metrics over traditional academic values. This transformation, according to Watermeyer (2019), has introduced a competitive accountability framework that places significant pressure on academics to demonstrate immediate social impact and public legitimacy, often at the expense of deeper, critical scholarship and long-term public engagement.

Watermeyer's critique extends to the increasing commodification of academic functions that are easily quantifiable and align with market principles, but often undermining the intrinsic values of education and critical intellectual engagement which will generally take far longer to transform from a theoretical position to one that might transfer into public knowledge. They question the prevailing trends of marketisation in higher education, highlighting the consequences these have on academic practice, governance, and the societal roles of educational institutions.

When specifically considering impact, Watermeyer and co-authors, in the book *The Impact Agenda: Controversies, Consequences and Challenges* (ibid.), outline how the demand for measurable impact on society and the economy is shaping the behaviour, culture, and ethos of academic researchers and institutions. They discuss both the positive and negative outcomes of this shift, noting that while it can drive innovation and societal benefits, it also pressures researchers to prioritise impact over scholarly quality and can lead to ethical dilemmas. And they argue that it may lead to instrumentalism (which mirrors aspects of neoliberal ideology is seeing the institution as a means to an end, rather than as a place of knowledge creation driven by academic curiosity), where choices about what to study are driven by what has the most potential for impact, rather than scientific inquiry or social need. When reviewing the policy drivers over recent years and the clear neoliberal influences on academia that have emerged from constant government reviews, it is difficult to disagree with this viewpoint. However, we would argue that there is much academic value in research impact if it is done not because of an assessment exercise or value for money policy arguments, but because academics can see its intrinsic value, we might get closer to a model where impact does matter to academics.

Conclusion

In conclusion, the challenge of integrating research impact into academic identity is deeply rooted in the perception that it is an externally imposed requirement, driven by neoliberal policies that prioritise market-driven metrics over intrinsic academic values. This top-down approach has led to resistance and a lack of genuine engagement with the impact agenda. Critics argue that the neoliberal transformation of higher education undermines critical thinking, intellectual curiosity, and the traditional role of universities as public goods. To foster meaningful impact, it is crucial to align impact initiatives with the intrinsic motivations and identities of academics, emphasising the importance of research in driving societal change rather than merely fulfilling external mandates. Addressing these issues requires a re-evaluation of how impact is framed and integrated into the academic culture, moving towards a model that values both scholarly excellence and societal contributions.

To foster meaningful research impact, it is essential to align these initiatives with the intrinsic motivations of academics. Emphasising the importance of research in driving societal change rather than merely fulfilling external mandates can help bridge this gap. This alignment requires a re-evaluation of how research impact is framed and integrated into academic culture, advocating for a model that values both scholarly excellence and societal contributions.

Moreover, addressing these issues involves understanding the historical context and evolution of higher education policies. The shift from liberal and civic intentions towards more neoliberal approaches has significantly influenced current academic practices and perceptions. Recognising this shift can inform strategies to better embed research impact into academic identity, making it a fundamental aspect of scholarly work rather than an external obligation.

Ultimately, fostering a genuine commitment to research impact requires institutions to support and motivate academics intrinsically. By valuing the broader societal contributions of research and creating an environment that encourages academic freedom and critical inquiry, universities can enhance their role in driving positive societal change while maintaining their core academic values.

Notes

1 www.ukri.org/who-we-are/research-england/research-excellence/ref-impact/ [Accessed June 2024].
2 https://civicuniversitynetwork.co.uk/ [Accessed June 2024].
3 https://upp-foundation.org/about-us/civic-university-network/ [Accessed June 2024].
4 https://kef.ac.uk/ [Accessed June 2024]
5 https://kef.ac.uk/about [Accessed June 2024]
6 https://sdgs.un.org/goals [Accessed June 2024].
7 www.un.org/en/academic-impact/universities-getting-ready-sdgs [Accessed June 2024].
8 www.ukri.org/what-we-do/browse-our-areas-of-investment-and-support/global-challenges-research-fund/ [Accessed June 2024].
9 https://education-uk.org/documents/dearing1997/dearing1997.html [Accessed June 2024].
10 www.ncub.co.uk/wp-content/uploads/2003/12/Lambert-Review.pdf [Accessed June 2024].
11 https://assets.publishing.service.gov.uk/media/5a7c9607ed915d12ab4bbc4e/0118404865.pdf [Accessed June 2024].
12 https://assets.publishing.service.gov.uk/media/5a7f289540f0b62305b856fc/bis-10-1208-securing-sustainable-higher-education-browne-report.pdf [Accessed June 2024].
13 www.researchinfonet.org/finch/ [Accessed June 2024].
14 https://assets.publishing.service.gov.uk/media/6409fda2d3bf7f02fef8832b/rdi-landscape-review.pdf [Accessed June 2024].
15 https://assets.publishing.service.gov.uk/media/5ceeb35740f0b62373577770/Review_of_post_18_education_and_funding.pdf [Accessed June 2024].

References

Barnett, R. (1994). *The limits of competence: Knowledge, higher education and society.* Open University Press.

Barnett, R. (2016). Recovering the civic university. In MacLabhrainn, I. and McIlrath, L. (Eds.), *Higher education and civic engagement: International perspectives* (pp. 25–35). Routledge.

Brown, W. (2019). *In the ruins of neoliberalism: The rise of antidemocratic politics in the West.* Columbia University Press.

Busch, L. (2023). *Knowledge for sale: The neoliberal takeover of higher education*. MIT Press.

Cantor, N. (2020). The Urgency of Recommitting Higher Education to the Public Good in 2020 and Beyond. Unpublished Paper. Rutgers University–Newark.

Checkoway, B. (2001). Renewing the civic mission of the American research university. *The Journal of Higher Education*, *72*(2), 125–147.

Cooksey, D. (2006). *A review of UK health research funding*. HMSO

Curtis, R. (2011). *In search of RAE-ality: The UK research assessment exercise (RAE) and the politics of academic freedom*. The University of Manchester.

Dekker, T. J. (2020). Teaching critical thinking through engagement with multiplicity. *Thinking Skills and Creativity*, 37, 100701.

Dilling, L., & Lemos, M. C. (2011). Creating usable science: Opportunities and constraints for climate knowledge use and their implications for science policy. *Global Environmental Change*, *21*(2), 680–689.

Donner, W., Urbinati, N., & Zakaras, A. (2007). John Stuart Mill on education and democracy. In Urbinati, N. and Zakaras, A. (Eds.), *JS Mill's political thought. A bicentennial reassessment* (pp. 250–274). Cambridge University Press.

Edwards, M. A., & Roy, S. (2017). Academic research in the 21st century: Maintaining scientific integrity in a climate of perverse incentives and hypercompetition. *Environmental Engineering Science*, *34*(1), 51–61.

Foskett, N. (2010). Markets, government, funding and the marketisation of UK higher education. In Molesworth, M., Scullion, R., and Nixon, E. (Eds.), *The marketisation of higher education and the student as consumer* (pp. 39–52). Routledge.

Furedi, F. (2010). Introduction to the marketisation of higher education and the student as consumer. In Molesworth, M., Scullion, R., and Nixon, E. (Eds.), *The marketisation of higher education and the student as consumer* (pp. 15–22). Routledge.

Giroux, H. (2002). Neoliberalism, corporate culture, and the promise of higher education: The university as a democratic public sphere. *Harvard Educational Review*, *72*(4), 425–464.

Giroux, H. A. (2010). Bare pedagogy and the scourge of neoliberalism: Rethinking higher education as a democratic public sphere. *The Educational Forum*, *74*(3), 184–196.

Harris, S. (2005). Rethinking academic identities in neo-liberal times. *Teaching in Higher Education*, *10*(4), 421–433.

Jones, D. S. (2014). *Masters of the Universe: Hayek, Friedman, and the birth of neoliberal politics-updated edition*. Princeton University Press.

Lynch, K. (2006). Neo-liberalism and marketisation: The implications for higher education. *European Educational Research Journal*, *5*(1), 1–17.

Marginson, S. (2011). Higher education and public good. *Higher Education Quarterly*, *65*(4), 411–433.

Mill, J. S. (1867). *Inaugural address: Delivered to the University of St. Andrews*. Longmans, Green, Reader, and Dyer.

Molesworth, M., Scullion, R., & Nixon, E. (Eds.). (2011). *The marketisation of higher education and the student as consumer*. Routledge.

Okolie, U. C., Igwe, P. A., Nwosu, H. E., Eneje, B. C., & Mlanga, S. (2020). Enhancing graduate employability: Why do higher education institutions have problems with teaching generic skills?. *Policy Futures in Education*, *18*(2), 294–313.

Olssen, M. (2021). Neoliberal competition in higher education today: Research, accountability and impact. In Olssen, M. (Ed.), *A normative Foucauldian* (pp. 307–327). Brill.

Olssen, M., & Peters, M. A. (2005). Neoliberalism, higher education and the knowledge economy: From the free market to knowledge capitalism. *Journal of Education Policy, 20*(3), 313–345.

Ostrander, S. A. (2004). Democracy, civic participation, and the university: A comparative study of civic engagement on five campuses. *Nonprofit and Voluntary Sector Quarterly, 33*(1), 74–93.

Pusser, B. (2006). Reconsidering higher education and the public good. In Tierney, W. (Ed.), *Governance and the Public Good* (pp. 11–28). Suny Press.

Radder, H. (2010). *The commodification of academic research. Science and the modern university.* University of Pittsburgh Press.

Smith, K. E., Bandola-Gill, J., Meer, N., Stewart, E., & Watermeyer, R. (2020). *The impact agenda: Controversies, consequences and challenges.* Policy Press.

Stevens, A. (2020). Governments cannot just 'follow the science' on COVID-19. *Nature Human Behaviour, 4*(6), 560–560.

Thornton, M. (2009). Academic un-freedom in the new knowledge economy. In Angela Brew & Lisa Lucas (Eds.), *Academic Research and Researchers* (pp. 19–34). Oxford University Press.

Tomlinson, M. (2017). Forms of graduate capital and their relationship to graduate employability. *Education+ Training, 59*(4), 338–352.

UPP Foundation. (2019). Truly Civic Report. https://upp-foundation.org/wp-content/uploads/2019/02/Civic-University-Commission-Final-Report.pdf [Accessed June 2024].

Ward, L. (2023). John Stuart Mill on the political significance of higher education. *History of Education Quarterly, 63*(3), 336–356.

Watermeyer, R. (2019). *Competitive accountability in academic life: The struggle for social impact and public legitimacy.* Edward Elgar Publishing.

Watermeyer, R., & Tomlinson, M. (2018). The marketization of pedagogy and the problem of 'competitive accountability'. In Watermeyer, R., & Tomlinson, M. (Eds.), *Pedagogical peculiarities* (pp. 87–98). Brill.

Williams, G. (2016). Higher education: Public good or private commodity?. *London Review of Education, 14*(1), 131–142.

4 Research Policy and the Problem for Institutions

Introduction

As most reading this book probably know, there is no such thing as a single policy for research impact. As is the case documented with research policy (Chandler et al. 2021; Thrift 2022), or perhaps where it is more specifically known as science policy (British Academy 2019), it exists in a myriad of policies which influence how higher education institutions are funded and apparently must respond accordingly. As we outlined in the previous chapter, Smith et al. (2020) define a cluster of impositions on the sector which they refer to as the "impact agenda"; a series of mechanisms around which universities must adjust their behaviour to and be assessed against. In considering the question of policy here, we address both the question of what this policy agenda comprises as well as the role of institutions in interpreting and enacting it. It is in the relationship between the two that we find both the challenge and the opportunity, and challenges and opportunities that are constantly remaking themselves with ever new policy intervention.

The Impact Agenda

There are a range of broadly neoliberal reasons why universities might now be more interested in impact than previously, and which aren't as a direct consequence of specific policies which require the sector to demonstrate the value of research through the impact it produces, but more indirectly because of the neoliberalisation of the sector. There is a kind of triple bind of neoliberalism which creates the conditions for the particular manifestation of impact that we will later discuss seeing in institutions.

First, and prominent within the policy landscape for impact, there is an increasing pervasiveness of the "innovation" framing discussions of research. Innovation as a term is now used in UK policy as part of the phrasing of "research" (UK Research and Innovation, UKRI),[1] and in this context refers to activities which bring ideas to market, that is, a set of activities that are either directly income generating or deliver a broadly economically beneficial or commercial set of activities (Kline and Rosenberg 2010; Tuckerman et al.

DOI: 10.4324/9781003454991-4

2023). Perren and Sapsed (2013) demonstrate the growth of the use of the word "innovation" in UK politics from 1960 to 2005, and this seems to be increasingly growing, not only in policies for higher education, but also in structures and strategies for funding.

This relates fundamentally to the second strand of neoliberalism in relation to impact, the emergence of impact "opportunity", and the idea that the higher education sector must support innovation to solutions where the state may previously have responded to a problem with direct funding from public money; this relates both to efficiency of use of public money but also to the significant challenges caused by the reduction in state support. Although this has been analysed as a significant policy discourse for several decades (e.g. Larner 2003), there is strong evidence to suggest the increasing prevalence of this across all sectors, including areas such as healthcare, social care, education, and legal aid, to name a few, where those in institutions are seeking to respond to the gap this rolling back of the state creates using their own skills and expertise. Snowball, Tarentaal, and Sapsed (2022), for example, highlight how there is greater emphasis on higher education institution (HEI) involvement in driving market-based solutions to sectors such as the cultural and creative industries which have previously been recipients of greater financial subsidy.

Commercial application of research ideas isn't new, but what is new is that it is now trying to find ways to embed it in new areas and new ways. This means that universities are increasingly seeing policies relating to innovation as a way to generate research funding for their institutions, the specific discourse of which is then increasingly prevalent around discussions of research impact. We discuss this later in that they are part of impact policy (through impact activity rather than impact demonstration and evidencing, as exemplified in the REF), but this shifts the way in which funding for institutions can be accessed.

Finally, broadly, that impact stories may be closely linked to student recruitment where institutions must sell their wares and what is most exciting about their institutions to those who are choosing where to study for a degree; those involved in university marketing seek out the kind of good news stories where researchers from their institution have made ground-breaking discoveries that are relevant to society. Indeed, it could be said that this kind of promotion by universities of their work has appeared well in advance of the impact agenda, as a consequence of these institutions being part of a higher education market where being seen to be doing interesting and "relevant" research, communicated in lay terms, is important.

We note that the inclusion of "research culture" and "rigour" (which is currently translated by some within the sector to mean ethics[2]) within recent policy rhetoric and soon assessment (in REF2029) is potentially an attempt to address any rampant takeover of market-based innovation, but taken as a whole this creates a complex set of frameworks and funding against which UK higher education is asked to assess itself as a sector, and here we look at the way in which this manifests at the level of the institution.

Clearly there are further market-related reasons as to why institutions have an imperative to be interested in having impactful research, such as to give profile and presence for the institution on a whole range of stages where they might seek to show off their expertise and specialisms. Stories of research that people can connect to are hugely valuable for that type of remit. There are also potential reputational gains for the institution in having researchers who can engage beyond the academy, such as through public engagement work which may assist an institution at a more local level particularly in helping decision-makers and voters to view the institution as a valued part of place, and therefore being supportive of, for example building projects and to be more positively receptive (or less hostile) to students. These aren't themes we develop further here because they've have been written about elsewhere, but that there are myriad ways in which impact is indirectly valuable to other consequences of the neoliberalisation of the sector and the consequential need for universities to protect their interests.

In this chapter we will outline the current policies that are relevant for shaping impact in UK institutions. We outline the broad policy context through which institutions are tasked with enabling impact and how this is growing to further prioritise pathways to impact based in neoliberal approaches. We argue that, despite REF not being the only policy mechanism, the inclusion of impact within this framework has been fundamental for shaping the position of impact within institutions. We suggest that although impact often doesn't play out in this neoliberal way, with analysis of impact case studies demonstrating the plurality of approaches, the centrality of REF to impact creates a culture of institutional competitiveness. Then we will look at how institutions are currently responding to this agenda as is evident in the data we've gathered, and the conclusions about the problem institutions have in relation to it.

UK National Policy on Research Impact

Given there is no single "research impact" policy in the United Kingdom, we will start by identifying some of the policies and policy mechanisms through which impact is manifest, encouraged, and promoted in academia. Existing work appears to be primarily about research funding policy (British Academy 2019; Thrift 2022), or REF specifically (Boeren 2023; Kelly 2016, 2022; Pinar and Horne 2022), and we seek to highlight here how understanding impact in the context of policy mechanisms which include these elements is more complex than a single context. There is a fuller discourse based around a whole set of policy-based mechanisms rather than a mandate, and therefore, in theory, institutions have the freedom to respond to the impact agenda however they choose.

This is not to detract from the influence the REF has had within the sector, particularly for embedding impact as an individual "narrative" , and particularly (as we discussed in previous chapters) as opposed to the collection of metrics or a narrative for the institution, such as with KEF or Civic University

Agreements. There's no doubt that REF2014, which brought in funding directly related to the quality of the institution's research impact, was the fundamental point at which impact became something that universities were specifically interested in driving and delivering. As we will discuss later, we argue that the way in which the relationship between funding policy and academia is enacted is part of the problem for impact, but we will start by outlining REF as well as other policy-related instruments relating to research impact, providing an overview of a fuller range of these instruments and the role they aim to play in embedding impact within the UK research ecosystem.

In the United Kingdom, key impact policy mechanisms clearly exist under the umbrella of UKRI, a non-departmental public body sponsored by the Department for Science, Innovation and Technology (DSIT) which, for 2024–2025, has had a budget allocation of almost £8.9 billion (or £25 billion over three years to 2025) (UKRI 2022a). This budget is allocated to research and innovation activity through the nine bodies within it – seven research councils with different disciplinary remits, as well as Research England and Innovate UK. The way in which this funding is invested is currently guided by the *UKRI Strategy 2022–2027: Transforming Tomorrow Together* (UKRI 2022b), which is then translated into each of the nine bodies through their own strategic delivery plan. The UKRI strategy immediately demonstrates the positioning of impact as central to their ambition, with the foreword asserting that

> the rate of discovery and technological advance is astonishing, with unprecedented opportunities to create value for society and the economy...At the same time, there are major challenges as we tackle the unintended consequences of the first industrial revolution...These are complex and demanding challenges, but the new industrial revolution presents major opportunities to address them and to enrich and improve lives in the UK and around the world.

So one of the fundamental UK research strategies is to make the most of research through innovation and technology for economy and society, as well as addressing significant challenges, with research and/through innovation.

The funding delivered through UKRI is allocated to a range of people and bodies, including researchers, research institutions, of which many are HEIs but also includes independent research organisations, as well as funding for businesses, NGOs, the NHS, and others. This funding is allocated via a number of key mechanisms and forms of assessment, and it is in these mechanisms that the research impact is funded. There are two strands to this research funding allocation in the United Kingdom, as part of what is known in the United Kingdom as the dual support system. One of these is mainstream quality-related (QR), allocated via the Research Excellence Framework (REF); at the time of writing, the REF2029 has been announced as a future exercise, and initial decisions published (Research England, July 2023), but much of the detail on the "rules" is yet to be confirmed, having consulted with the sector about significant elements of this. As with the two previous REF exercises, though (REF2014 and REF2021), it seems impact is still firmly positioned within it, though renamed in scope to be engagement and impact, and to account for 25% weighting in the quality profile (with 50% for research outputs and 25% for people, culture, and environment). Watermeyer (2019) in particular has critiqued in detail the audit culture of impact, most notably visible in the form of REF impact case studies.

The second strand of the dual support system has delivered other significant government funding initiatives designed to stimulate impact in particular ways, including a significant proportion of funding allocated through what is now known as the new cross-UKRI strategic programmes, described as taking "learning from existing programmes such as the Industrial Strategy Challenge Fund, Fund for International Collaboration, Strength in Places Fund and the Strategic Priorities Fund" (UKRI funding overview). Although not new, for example Innovate UK having funded an ongoing mechanism for impact – knowledge transfer partnerships (KTPs) – for over 50 years, there has arguably been a strengthening of the policy position on impact. This strengthening is both in terms of the integration of innovation with research through the creation of UKRI in 2018, bringing together the Research Councils (RCUK) with Innovate UK under one strategic umbrella, as well as in the creation of specific impact funding, which we will discuss later. Opportunities for driving research and innovation, particularly in place, also comes through to institutions in the form of, for example, the UKRI connecting capabilities research enterprise development funding. In total, this means that a significant proportion of government's second research funding stream, via the research councils, is actually for impact and innovation, generally via relatively prescribed routes rather than open rounds of funding.

Impact in the Early Days

In the context of understanding both the influence of particular policy and associated mechanisms on the establishment of impact, as well the continuing legacy of how the shifting focus to impact was received within academia, it is important to consider more about the policies that brought it about and the relationship between these, HEIs, and academic practices. This is particularly important given the influence of a range of policies relating to research impact, which all appear to hold slightly different positions within higher education discourse.

It's well-documented that the inclusion of impact in REF2014 was a seminal policy moment for the 'impact agenda' (Smith et al. 2020). It came about after more than a decade of build-up (as outlined, e.g., by Williams and Grant 2018) whereby the government and its agencies, in conversation with the sector, considered options for appropriate assessment of research impact in HEIs. Williams and Grant (2018) note it wasn't the first instance of considering benefits of scientific research accruing to society at large, but the announcement of impact in REF2014 in 2011 (REF 2011) was the first moment universities had to actually do something about it rather than simply be part of the discussion around it.

At the time there was considerable vigorous argument put forward by academics of the value of their research to the world. The general consensus throughout the first decade of the millennium has been of academics feeling under attack by government policy, particularly it seems of the arts, humanities, and social sciences, through funding models for both degrees in these disciplines, and in other ways, such as the more general reduction in support for the arts (e.g. school music) and access to arts across the general population. Within this, there is a sense in which public understanding of and support for these disciplines has been diminished, and funding reduced, for example the reduction in funding for arts-based degrees and alongside that the reduction in degree programmes for humanities and arts-based degrees, and the downsizing or closure of departments in some institutions (whether for reasons entirely to do with viability or more to do with the ability of these degrees to pay high percentages to the "centre" of an institution). Within this discourse academics within these disciplines have argued for the value of impact in making the case for the role they play in society and for why they are valuable (e.g. Daybell 2019). Yet, there are many others who argue for the value of research as such with no government-driven agenda for impact attached to it, such as the union response to the inclusion of impact in the REF2014, signed by over 17,500 academics, higher education professionals, and researchers (UCU 2009).

The research councils were also central to the messaging around impact, if not necessarily initially to the implementation of impact. As government funders of original research, the research councils first introduced impact-specific requirements for funding bids in 2007 in the form of an "Impact Summary" and a "Pathways to Impact" section. Researchers were required to complete these as part of their application for research funding until they were abolished in 2020, not because impact was no longer important but perhaps because it was deemed to have reached maturity and would be embedded into the case for support of the research (Wilsdon 2020). These funding councils also developed specific schemes to fund impact, such as the AHRC's "follow-on funding" scheme, and also innovated about how to fund impact, ultimately devolving a lot of the responsibility to institutions deemed to have sufficient opportunity to generate impact from their research through block-grant funding in the form of Impact Acceleration Accounts.

The Stern Review of the initial REF – REF2014 – published in 2016, further cements the idea that impact should be a core element of academic practice across the sector. As Stern writes (DBIS 2016: 22): "whilst the nature of impact will vary greatly across disciplines and activities, it is in all academics' interests to be able to indicate the impact of the research that they undertake". This, he implies, is separate to audit though, with case studies being an institutional concern, and one that is about seeking to get closer to a position of highlighting the best kinds of impact from research:

> Therefore it is important that each Unit of Assessment is required to submit case studies. However, there are a number of reasons to believe that it would be beneficial to relax the tight coupling between the number of staff submitted to a Unit of Assessment and the number of case studies required, and give institutions greater flexibility to submit case studies which may cross the boundaries of different units. Giving institutions some flexibility in the distribution of their case studies would allow them to demonstrate their strengths more effectively, and make it easier to submit strongly inter-disciplinary case studies.

This offers firm leadership in this area, with Stern himself having clear evidence of his own practice in this space which pre-dates any government audit of impact.

Policy Shifts Around How Impact Pathways Should Be Established and Conceived of

Much of the use of funding that's geared towards impact is either "challenge-led" funding – that is it is themed in such a way that it encourages research and collaboration to establish new ways of addressing core issues in society, for example an ageing population, the climate crisis, sustainability, cybersecurity etc., or it is framed through existing, sector-wide "pathways" to impact, as is evident in the targeting of funding to universities in the form of policy QR, participatory research QR (in very small amounts), and commercialisation such as through various commercial pre-accelerators and accelerators (e.g. the iCURe Programme[3] and now the SHAPE – Social Sciences, Humanities and Arts for People and the Economy equivalent ARC Accelerator[4]), as well as Higher Education Innovation Funding (HEIF). From a SHAPE perspective, there has been a significantly increased emphasis on commercial pathways to impact both through bespoke programmes such as the ASPECT Network (initially funded by Research England[5]) alongside the ARC Accelerator, and through increased emphasis on commercialisation as a pathway in the allocation of impact acceleration account funding; both the AHRC and ESRC have been presenting within the impact community about their plans around commercialisation. There is a sense from the funding councils that this is a very unknown and experimental space, with examples of research being turned into commercial ventures provided. The ethics and efficacy of this approach, except in a small number of cases, is as yet unproven though. For example, there is little consideration of whether other pathways might have been more suitable. Institutions are encouraged to use responsible innovation frameworks, but ultimately it is up to institutions to define the ethical framework around this activity.

Since 2020, Research England has also allocated an element of QR to HEIs for policy engagement from research, as well as an allocation of QR funding for participatory research. As a consequence of this quite small amount of funding, policy has suddenly become a hot topic for institutions, with policy roles and units appearing in many research and impact support directorates in universities, with other indicators of the growth of activity in this area being visible in, for example, the huge growth of the Universities Policy Engagement Network (UPEN); from "set up in 2018 by a small group of universities. It now has more than 100 UK university and policy relevant organisations as members" (UPEN 2023 Annual Report), which constitutes two-thirds of all UK universities joining this policy organisation. It has shifted from individual academics carving their own way through policy engagement from their research, to institutions putting in infrastructure to try to stimulate this activity.

A third strand of discourse around impact in recent times is about place and the role of institutions in places. In terms of policy strategy, as set out in *UKRI Strategy 2022–2027 Transforming Tomorrow Together*, this is almost exclusively about the role of research and innovation in regional economic growth, framed in terms of UK competitive advantage. In their second of six objectives, *World-class Places*, UKRI outline that is about "securing the UK's position as a globally leading research and innovation nation with outstanding institutions, infrastructures, sectors and clusters…including in areas of low R&D intensity" (UKRI 2022b: 18–20). This connection to the government levelling-up agenda, with a fundamental link to the concept of place, is principally framed here in terms of economic growth and development. Despite suggesting that "UKRI will deliver economic, social, and cultural benefits from research and innovation to all UK citizens" it goes on to outline the main value it sees as being to "address regional disparities and enhance place-based livelihoods and economies". This therefore seems to view place as a way of clustering particular kinds of economic activity and sector around research expertise (people) and facilities to drive economic growth (then with knock on effects for society and culture).

This is somewhat in contrast to the version of place and the local that is highlighted in the Rand report of the REF2021 impact case studies (Stevenson et al. 2023), which examined the idea of hyperlocal impacts. By this they developed the idea of hyperlocal as anything within a 25-mile radius of an institution, and identified these in impact case studies from the submitting institutions. As part of this they took Manchester as an example of a place with significant amounts of hyperlocal impact, which ultimately, they suggested, was as a consequence of the highly participatory methodologies of institutions local to Manchester leading to a huge range of different types of impact, from cultural to economic, but with a much broader and more holistic idea of the role of the university as an institution within these places. This is not to say that this concept is without challenges. For example, there are more than 10 times the number of people within a 25-mile radius of Manchester universities than in more peripheral locations such as Norwich, Aberystwyth, or indeed somewhere like the University of Highland and Islands with multiple campuses across a wide and mostly sparsely populated area. There are also significantly fewer universities clustered in these areas, so the "place" of these universities perhaps stretches far beyond a 25-mile radius (and this also makes the idea of capturing hyperlocal impacts which demonstrate reach and significance probably more problematic to these institutions), but this concept nonetheless helps to demonstrate the plurality of place-based research impact. It captures a liveness of place and of human value, and demonstrates what is perhaps closer to the civic idea of universities than the UKRI strategic view.

It has recently been noted though that there's little conceptualisation of the role of universities as civic institutions to research and impact (e.g. Dobson and Ferrari 2023). It is often wrapped up in the general discussion about civic universities, and implied as something that is done or could be done. Within a framework developed to support the Civic University Network (the main body supporting institutions to develop civic agreements), seven domains of civic impact have been developed. Core aspects of universities such as their facilities, infrastructure, education, and students are all mentioned within this. Research is alluded to in elements such as "How can our university help our places move from 'functioning' to 'flourishing'?" and " How will [the university] influence environmental behaviours throughout our city of region?" in relation to *Environment, Climate, and Biodiversity*, but the only explicit mention of research within the high-level framework is "What would it look like if our civic priorities were embedded throughout our core activities of teaching, learning, and research?", as a question around institutional strategy and leadership. The report underpinning it (Civic University Network n.d.) does also mention research, but again bundled up with teaching and learning.

Some have written about examples of civic action through community-based research, principally identifying best practice for how to engage with community partners and communities, but without as much consideration of the broader academic context around it. For example, Jarvis (2023) strongly advocates for the value of a "City Conversations" project, with best practice demonstrated through training and implementation of community researchers within a particularly deprived area within Liverpool. It cites success factors, including building relationships and trust with community and public sector stakeholders, project management, and relationship mechanisms from the outset, and a continued commitment to co-production methodologies. It then concludes that the project was

> time limited to the research grant period. This is a typical scenario for many university–community research projects, which stop once grant funding ends despite demonstrating social impact. Ensuring long-term sustainability for public engagement and research with marginalised groups is a live issue for universities.
>
> (Jarvis 2023: 47)

This is no-doubt well-observed in terms of the challenge of sustainability for this type of work across many universities, but the appearance of a long-term commitment to sustainable funding for universities to do this seems unlikely. Clearly universities could commit part of their QR and HEIF grants to this work, and there are examples of these funding sources being used for this type of community work (and the KEF enables institutions to highlight this in their narrative), however there seem to be plenty of examples of these kinds of, no doubt, valuable projects lamenting the lack of long-term funding, and very little evidence of long-term funding forthcoming.

This is not to dismiss valuable work that's being done in this space by universities, but to raise the question about where research impact "fits" into this model, and to suggest that it doesn't currently fit within it very well. It could simply be that the answer is that it's organic and looks for opportunities, which is one possibility, but we argue first that the current broad-brush approach to including research (and by default, impact) within civic discourse means that it sits quite uneasily within this, with many challenges remaining; and second, that it's only by broadening an understanding of the value of research within civic universities away from quite narrow definitions of locality and place that it is able to find a more useful role within this ecosystem.

Narrow definitions also means that funding is increasingly restrictive, with funders clearly seeking to drive impact in very instrumental ways, and not necessarily always apparently very coherently around the different thematic strands developing in impact discourse; for example in one UKRI funding stream that sought local, commercial projects with sustainability impact, all in one short application.The combination of it being a requirement to have a locally oriented project with a focus on commercialisation and sustainability (with commercial tending to require a larger than local market for success, and sustainability having potential for global implications), is quite an odd and difficult mix of scales to fit entirely within one place, particularly somewhere with a smaller population and fewer commercial opportunities.

Approaches to Impact in Institutions

As we have discussed, government strategies relating to impact have created a significant question for institutions about how to ensure their researchers understand what they need to do. As a discourse, impact has led to the emergence of a new way of thinking for these institutions, despite related terms such as knowledge transfer and public engagement already being in use prior to the impact agenda. Impact emerged as something that was somewhat amorphous, yet acutely relevant across the whole institution with increasing relevance across a whole range of policy mechanisms. Guidance was produced by research directorates and funding support functions about how to fill in the impact sections of grant bids, advising people about what sort of things they needed to think about and how to generate the sort of connections they might need for these sections, but at the same time with very little evidence about what was actually reviewed well, or indeed very little evidence that these sections really mattered. After all, funding bids were still reviewed by subject peers with track-records for research. Rumour abounded that some councils occasionally used these sections as a "tie-breaker" where they couldn't fund all top-scoring research, but with little consistency across councils. Institutions, on the other hand, were also concerned with how to access funding for impact and ensure impact-related activities were possible, and it was the Pathways to Impact section that offered this opportunity. The problem of researchers carrying out their planned research and then approaching their institution for funding for the softer non-research activities, like engagement events with those in a range of stakeholder roles, or producing reports for specific non-academic audiences, and even having specific non-research roles and access to expertise, could be partly solved with Pathways to Impact, with institutions expecting funding to be requested for this (e.g. Manchester),[6] if only everyone knew how to do this.

In reality, we suggest that this key element wasn't especially comprehensively supported. Certainly, there was guidance produced within institutions at the time (much of which is still available even four years after the end of Pathways to Impact), often with an overview of what was being asked for in each of the two impact sections, and "top tips" type advice about how to go about filling them in. This advice often revolved around public engagement with research principles such as to engage potential partners early in the process, suggestions of the types of activity that might be appropriate, and the direction to cost it into the bid. How was this guidance rolled out though? Funding bids are typically supported to varying degrees by research-funding advice teams – those who tend to have knowledge of research funding and funders, and of what makes a good bid. Impact was new though, so it is not clear if and how these teams were immediately upskilled to be able to support impact in bids. In the early days of impact we assume that there were very few (if any) impact managers, but knowledge exchange professionals – those with a remit for things like intellectual property (IP), public engagement, and strategic partnerships – may have been asked for advice on these sections, probably in a very specific ways (e.g. "do you have any contacts in x sector"). The sections would certainly have been talked about at grant-writing training, though the audience for these would mostly be early career researchers and so upskilling the full population of researchers was unlikely. For institutions though, this was probably not an issue. After all, these sections on bid were of very little significance (indicated by some academics to be completely useless).

Next came REF2014 though, and as we discuss through examples of our own experience in Chapter 1, we saw little in the way of preparation in the years preceding. There were suggestions of limited attempts to ask people to keep records about their impact, but there wasn't a sector-level solution to any of these things. It was the very early days of the impact professional and of the impact industry, with various similar kinds of role (e.g. public engagement, knowledge exchange, and evaluation professionals) being asked to take on some of the impact work. Other institutions it seems did create new (and certainly anecdotally, only with temporary contracts) impact-focused roles in combination with using a range of consultants from specialisms such as science communication and other kinds of communication backgrounds to enhance their articulation of impact in the form of a case study. In effect, institutions were seeking to enhance what Watermeyer (2019: 73) refers as the "charm offensive", but arguably this is really because institutions were trying to find some kind of location of the skills needed to write an impact case study. On the converse side of this were the institutions where academics were asked to write an impact case study themselves. The point here is that institutions were in quite an unknown position, without a particularly clear idea of strategic alignment, support, or professionalisation, and in effect, all solutions in use in institutions were experimental for a really quite unknown end.

Clearly this position has changed significantly and through FOI data from higher education institutions and ethnographic research we explore the way in which institutions have now responded to impact in UK higher education policy. In total we received responses from 108 institutions, representing over two-thirds of the 157 higher education institutions that made a submission to the REF.[7] We present a summary of the responses further. The intention here is not to provide a detailed statistical analysis of each response but to look at the sector as a whole in terms of practice around research impact in institutions. Again, this is an opportunity to test our view that the sector as a whole, while responding to the needs of research impact as a measure of research assessment (and therefore long-term research funding), has recently still approached impact in an ad hoc manner, with much reliance on external consultants rather than embedded knowledge.

Furthermore, as we have argued in previous chapters, if we have the impact agenda imposed by management and university leadership, rather than developed from the ground up by academics who see it as part of their identity and appreciate its fit with broader research culture, yet those who impose it show little evidence of appreciating its value beyond a metric to submit as part of research assessment, it seems unlikely that cultural change will be significant.

As we outline in Chapter 1, we asked a series of questions designed to gauge the position of research impact and level of practical support for impact both within institutions and as coming from outside institutions, and we will discuss the outcomes here.

FOI requests allow scope for institutions to refuse to respond to any or all questions for a limited number of reasons, which we will also consider in the context of what it means for how institutions position impact. A "Section 43 (2)" refusal relates to a specific (sub)section in the UK's Freedom of Information Act 2000, the legal mechanism under which we had issued the FOI requests. This section in the legislation is used to exempt information from being disclosed if its release could, or would be likely to, prejudice the commercial interests of any person or organisation, including the public authority holding it. In general, the term "commercial interests" refers to an information-provider's ability to successfully participate in a commercial activity, such as contract negotiations, competitive bidding, or other business transactions where exposure of certain information could disadvantage one's competitive position.

This exemption is subject to a public interest test, meaning that even if the information might prejudice commercial interests, it must still be disclosed if the public interest in releasing the information outweighs the public interest in withholding it. Therefore, we would assume that any university deciding to use this mechanism in the legislation to refuse our request would have already considered the public interest in the information and decided that this is outweighed by the commercial risk in exposing the information to a public setting.

We also, in a few responses (including one institution who used this for their whole response), received a number refusing responses based upon a Section 12 (1) refusal. A Section 12 refusal under the UK's Freedom of Information Act 2000 occurs when complying with a FOI request would exceed the cost limit set by the regulator. This means that if the time and resources needed to locate, retrieve, and extract the information exceed a certain financial threshold, the public authority can refuse to process the request further. Typically for an educational institution, this limit is set to around £450 or 18 hours of work.

When invoking Section 12, the authority does not need to confirm or deny whether the information exists, if even estimating this would exceed the cost threshold. However, one might assume if the institution was confident it held no information, it would know that it would not cost them anything to respond to the request. Therefore, we can assume that a Section 12 (1) refusal is more likely to result from the fact that this information is not readily to hand which, given the basic nature of these requests, would suggest that information related to impact policy and practice is not readily to hand.

Do you provide training for academics about research impact? If so, what is the nature of this training?

The majority of institutions said that they did provide some form of training for academics, with 99 saying they did. Only three institutions said they did not, and then three refused to answer based upon a "Section 43(2) refusal" and one based upon a "Section 12 (1) refusal". In a limited number of cases then it was either thought to be commercially sensitive to disclose this information, or that it would've taken in excess of 18 hours to compile this level of information.

The summaries provided regarding the nature of the training disclosed by 99 establishments detailed a large number of different approaches to impact training.

Impact training for academics was broad-ranging and delivered in various formats, including webinars, one-to-one sessions, and workshops, and were designed to address different aspects of research impact, such as evidencing, case study development, and the use of social media for research impact. However, not all institutions provided the same sort of training, and the level of detail about the nature of the training was also quite variable in response.

The University of Bath, for instance, provides a detailed and structured impact training programme for all academic levels, from basic introductions for new academics to fellowship programmes and personalised coaching and mentoring. These sessions cover a spectrum of impact-related activities, including planning, delivery, and evidence of impact, with resources like impact toolkits to support researchers. It was noted that in a number of cases respondents were keen to stress that training was open to individuals across different faculties, disciplines, and career stages, ensuring that training is relevant and accessible to all.

Moreover, the training often included sessions on broader aspects of what we might consider relating to research impact, such as commercialisation, entrepreneurship, public engagement, knowledge transfer, and communication, with, for example, specialised training from the Institute for Policy Research at the University of Bath providing one-to-one training for policy fellowship programmes and advice on engaging with policy audiences. Some universities tailored their programmes based on annual needs assessments as part of their REF cycles, including several universities who reported that they focus on specific training for the REF, with targeted webinars and workshops aimed at preparing academics for REF submissions.

Collaboration is key in these programmes, as evidenced by the involvement of various university departments and external providers, ensuring that the training is comprehensive and that the specific needs of different research disciplines are met. For example, the University of Cambridge offers a diverse set of activities supporting entrepreneurship and public engagement, and the University of Liverpool has an ongoing development programme to enhance the skills and knowledge of its researchers in impact creation. There was also a great deal of variability in who delivered the training, with some saying all training was delivered in-house, whereas others made a lot of use of external consultants.

Some respondents also said that they were planning new workshops and adapting to changes in the REF, aiming to demystify its impact component. Given, at the time of writing, the detailed guidance for REF2029 has not been published, it is interesting to see that REF training is still continuing with an expectation that it will be required with similar expectations to previous REF assessments, but specific training for REF2029 isn't yet possible.

Overall there was a great deal of variety in what was described as being delivered which is not a surprise given how different institutions have adapted to calls for impact management and assessment in different ways. It was interesting to note that, for the majority of institutions, delivery was done by either in-house professional service staff or external consultants. There were no universities who specifically said that impact training was delivered by academic staff. This is perhaps the most telling finding from the responses, and once again highlights the disconnect between research impact as part of academic identity and the view by some that impact is a management and policy imposition rather than something done by academics. Given we issued this request after the REF2021 results were published, it would be unusual for any institution to not have academics on their payroll that had not performed well in this assessment, yet what was reflected in practice descriptions was that they would not look to their internal academic staff to provide guidance to other academics on how to do impact effectively.

Do you provide training for professional service staff about research impact? If so, what is the nature of this training?

While there were still a majority (80) of respondents who said that they did deliver impact training to professional service staff, almost a quarter (25) said they did not. We also have three Section 43 (2) refusals and one Section 12 (1) refusal.

Again, in descriptions of what was provided, we could see delivery by both internal teams and external consultants, focusing on topics such as evaluating engagement, impact for the REF, and embedding impact into grant applications. Workshops and one-to-one training sessions are offered to a range of university staff, tailored to suit their career stage and the discipline(s) these staff are embedded within. Some said they offered the same training to all staff – to both academic and professionals – while some provided training specifically for their impact professionals. Many universities offer regular training, including in-person and online workshops, as part of their researcher development programmes often also open to professional services staff, and provided it in formats such as workshops, seminars, one-to-one sessions, online resources. It also included more specialised training for certain staff roles, particularly those involved in grant support management, and use of external providers and online training to deliver for this, as well as for more general introductions to research impact. These often include sessions on stakeholder analysis, data collection, and using software tools like Impact Tracker**. Some identified how the training varied from introductory to advanced levels, and many institutions also talked about provision to attend external workshops and conferences. The training appears to aim to cover the entire lifecycle of research impact and is adapted to the needs of various disciplines.

Many universities report also creating networks to develop skills in impact and knowledge exchange, with staff encouraged to engage in professional service networks such as those offered by organisations such as the Association of Research Managers and Administrators (ARMA), and also look to ARMA for the training of their impact professional staff.

Overall, the descriptions of the training centred on maximising the impact of research, ranging from stakeholder engagement and public policy to commercialisation and entrepreneurship, and making sure those who supported research academics would understand the research impact process. However, as we have observed earlier, this was something which did not involve academics, who might have had considerable impact knowledge, in training delivery. Also, given that the descriptions from academic training often talked about in-house professional services staff delivering that training, it is somewhat surprising to see the number of institutions who do not actually train their professional service staff.

How many FTE professional services staff do you have dedicated to research impact?

For this question, which was aimed at understanding whether there would be "impact specialists" on staff or whether impact was just viewed as a function of the wider research professional staff base, we saw a great deal of variability. First of all, we received three Section 43 (2) refusals and one Section 12 (1) refusal. Of those who did respond, 23 said they have no dedicated research impact staff. On average based upon all responses, institutions had just under 3 (2.86) FTE impact staff, with a standard deviation of 3.52, showing a reasonable spread across responses. The maximum number of impact staff returned by one respondent was 17.7 FTE, which was unsurprisingly from a large Russell Group institution. And while most of the responses with high FTE count were from research-intensive institutions, there were two post-92 institutions with an FTE count of over eight.

Furthermore, the actual number of impact professionals working in institutions as indicated in the FOI responses was 290 people (or FTE equivalent). As only 89 of 157 REF submitting institutions responded to this question, and several large Russell Group institutions, including the University of Oxford, the University of Birmingham, and Imperial College London, who stated they were unable to answer this question, and with several returning "0", where we know from networks and organisational websites that this is not the case, we estimate this figure to be significantly higher, and will likely grow in the build up to the REF2029. There are, therefore, hundreds of impact professionals working in universities in roles that, from our interviews and ethnography, it seems largely did not exist prior to the impact requirement in REF coming about in 2011. It seems then that the REF has brought about an entirely new type of research professional, and one which we go on to look at in more detail in Chapter 5, but that there currently seems to be a lack of understanding of within the sector – here we argue possibly as a consequence of both institutional oversight and in some cases perhaps because cross-institutional knowledge sharing is seen by some as detrimental to their competitive advantage.

We also note from cross-reference with interviewees in these institutions that the figures provided don't seem to include the other professionals supporting impact generation. Indeed, in several interviews we were told that knowledge exchange staff broadly (including, e.g. commercialisation, policy, and KTP support functions) are now organised and given purpose through "impact", but which haven't been reported in these figures.

Do you have a policy for the selection and development of impact case studies? If so, please can you provide it?

Further variability of practice comes from this question and the next one we will explore. We wanted to see whether there was a policy for selection of case studies to ensure transparency and a consistent approach to selection applied. This seemed a reasonable question given, in conversations with peers, there was always a senior management decision on the selection of case studies and what finally got submitted rather than basis in institutional policy. It would seem from our responses that it was rarely the case, with only 16 respondents saying they did have a policy. Those who said they did in most instances said that it was described in their REF Code of Practice, with a couple explicitly talking about a process of senior leaders, unit of assessment leads and research professionals making the decisions based upon evaluation criteria. However, this also means that for 84 institutions there was no uniform approach for how they did this, or at least nothing written and formalised that they could point to. This produces an inconsistency for REF, where institutions have to define their approach to selecting staff and outputs for submission to ensure equality and transparency, but not impact.

Is research impact a separate category for workload allocation for academics with significant responsibility for research?

We also asked a complimentary question, which was whether there was any specific workload allocation for those doing impact. While this is something of a contradictory question given our arguments that impact should be seen as a part of the research process, should the nature of the research lend itself to impact, rather than being treated as a separate function of research management. We asked this question to see more broadly whether research impact was a consideration in workload allocation, and particularly in light of us both having been involved in conversations with academics within and outside our own institutions where they were advocating that impact was difficult to do without a workload allocation for it. We were also mindful that a number of our interviewees, who had achieved high levels of esteem in at least one previous research assessment, had said that it was not something that had ever come up in their appraisal processes, so the idea that it might have been considered within workload doesn't seem likely.

In total, only five institutions indicated that impact was specified within workload allocations. Of those institutions with one, there is some indication that this is for those with impact leadership responsibilities. We also know of at least one institution who responded "no" where there is a workload allocation for those identified as impact case study leads, so again, this seems to demonstrate the lack of coordinated knowledge about impact within institutions. It is difficult to assess whether the lack of inclusion of impact within workload is because of a holistic approach to impact though, rather than a lack of consideration impact. If taken as a sign of embedding, this semblance of organisation towards impact is also visible in some of the ways in which academic posts are oriented, with several mechanisms coming up in our research, such as the inclusion of it in promotion criteria (which came up as a discussion point in many interviews), and as seen in job descriptions where it has become a standard part of the criteria for academic posts in some institutions. Both of these aspects were described in terms of a lack of consideration though, which indicates that this might be taken as an indication of a lack rather than embracing of impact. So the mechanisms institutions use seem to imply it is fundamental to academic roles but it's not necessarily actually embedded in the more active modes of formal support, such as appraisal and workload.

Do you have a civic university agreement? If so, please can you provide it?

Given the recent policy drive around civics, which do hark back to some of the origins of the discussions around public good and the value of universities to their communities, we thought it might also be interesting to ask how effectively this had influenced practice across the sector. From our responses, it would seem that the answer is not significant, with only 31 respondents having a civic university agreement in place. While this is slightly outside of the scope of this book, it is a useful reminder that policy driven nationally without the buy in of those tasked with implementing it is often doomed to fail (as has arguably been seen with the KEF).

How much did you spend on external consultants for research impact for REF2021?

Finally, we explore one of the more interesting questions we posed, which is again driven by our own observations and discussions with peers. It seemed to us that there was a lot of movement in the sector from those who proposed they could be commissioned by institutions to enhance impact performance in a given institution. Equally, we had seen institutions assuming that if they could bring in experts from outside of their institution, they would be able to improve things more than if they were to rely upon internal knowledge. This, in turn, made us suspect that many institutions did not have much confidence in their own abilities to support, assess, and articulate impact, and would prefer, instead, to rely on external consultants. Therefore, we wanted to see if we could measure the amount of (public) money that institutions have spent on external consultants over the previous REF period. This one question perhaps posed the most issues with respondents, and there was some discussion with FOI officers regarding why we wanted this information. In the end, we received responses from most institutions regarding this. Overall, 24 respondents said that they had not spent any money on external consultants, meaning the vast majority of institutions had.

We received 12 Section 43 (2) refusals, which we will explore in more detail further, and 5 Section 12 (1) refusals, which means that it would have taken the institution too long to discover how much money they had spent on this. This reflects our argument that some institutions are not approaching this area of research assessment with any consistency of practice or approach. Furthermore, seven institutions said that they did not know how much they had spent on consultants, again perhaps highlighting a lack of rigour in practice in managing research impact.

Nevertheless, in total we had 61 respondents who did attempt to place a figure on the amount they had spent on consultants during the previous REF. The smallest amount spent was £3000, the highest was £100,000. In total, our respondents said that they had spent £1,663,970.76 on external impact consultants in the previous REF. Given that we can assume those who said they don't know, didn't have time/resource to find out, or refused on the grounds of commercial sensitivity, all also spent some funds on external consultants (we would hope that those who said they didn't know had at least some semblance of knowledge they had spent more than £0), we would anticipate the actual figure spent on consultants in the sector overall is much higher than our calculated total. Which once again shows how reliant institutions are on those outside of their institutions to steer their impact policy and practice.

There are a number of issues arising from the FOI responses we will return to in the remainder of this chapter. However, there is one thing of particular interest, and an interesting by-product of the FOI data collection process, that is worth reflecting upon here if we are to consider the challenges of turning research policy into practice for institutions. It also, we would suggest, further highlights the marketisation mind-set that seems to have gripped many university leaders and is perhaps a reflection on what some authors have referred to as the neoliberalisation of the higher education sector (Maisuria and Cole 2017). A number of institutions turned down requests because Section 43 (2) of the Freedom of Information Act 2000. The literal wording of this subsection is:

Information is exempt information if its disclosure under this Act would, or would be likely to, prejudice the commercial interests of any person (including the public authority holding it).

While we did not have the time, nor inclination, to debate with institutions regarding whether this exemption was valid, we do think it is a worthwhile illustration that respondents believed that by divulging general detail about training they deliver related to research impact or how much they spent on research impact was in some way information that could be commercially exploited by "competitors". This is against a research assessment framework which judges each piece of research, let alone institution, upon its merits as evaluated by a community of peers.

While, clearly, there is funding related to a strong research assessment, it does not place institutions in competition with each other and it seems curious, and extremely misguided, to suggest that the information we requested would prejudice commercial interests. Were our respondents saying that if we were to discover how much they had spent on consultants during the last REF, we would be able to gain some form of competitive advantage over them? The suggestion here is that good research impact is something you can pay for, and if your "competition" spends the same, they too will have the same outcomes. Therefore, they view this information as commercially sensitive. Which does, we would suggest, strongly illustrate how disconnected from the underpinning research and academic practice impact has become in the eyes of some senior leaders in the higher education sector. It suggests there is a perception by respondents that impact is something that can be done with a technique or a process model, rather than hard work, relationships with stakeholders, and high-quality research.

In bringing this discussion to a close, we would suggest that the FOI results are another strong illustration of how inconsistently and, in some cases, ad hoc, policy and practice is applied to research impact by institutions. Training is given but not by impactful academics in settings. Research professionals are expected to deliver training but not all receive training themselves. Impact is rarely part of an academic's appraisal, and there is unlikely to be any workload allocation given to those who can demonstrate impact potential (even if the appraiser has the knowledge to assess this); policies to select case studies are generally not in place. Regardless of our views on the relationship between research and impact, there is clearly quite a dissonance between how institutions aim to make impact business as usual and the way in which academics experience this at an individual level. And institutions clearly do invest in impact, most visibly here in the form of training, professionals, and external consultants, but the lack of coherence of approach and lack of embedded knowledge, or at least a reluctance to draw on the embedded knowledge – as we will discuss in the two chapters that follow – means that institutions have perhaps responded quite reactively and superficially, rather than creating deep strategy to fully support research impact. It may appear "embedded", but we argue crucially not necessarily in terms of coherence of approach, and perhaps often not in configurations that entirely make sense to research.

We are, perhaps as a result of the extensive policy work we have, ourselves, done over the years, reminded of a satirical device often referred to as the politician's syllogism (Chen 2007). This is a term often used to explore the flawed logic of a policy decision that does not seem to have been informed by evidence and follows the following formulation:

1 We must do something.
2 This is something.
3 Therefore, we must do this.

The logical flaw here is that just because an action can be taken, it doesn't necessarily follow that it should be taken, especially without considering its implications, alternatives, or effectiveness. The term highlights how policy that is reactionary or driven by the desire to satisfy demands for action, regardless of whether the actions taken are genuinely beneficial or the best solution to a problem is rarely effective.

We do not, when we explore this data, have confidence that this is a sector that understands the nature of research impact in depth, and that those making research policy in institutions are unlikely to have those decisions informed by a deep knowledge. Which may, of itself, go some way to explaining why research impact, and those doing it, often feel disconnected from their institutions.

Conclusions

In conclusion, this chapter underscores the multifaceted nature of research impact policies and the complex relationship institutions have with these policies. The evolving policy landscape, heavily influenced by neoliberal ideologies, requires universities to navigate a myriad of expectations and funding mechanisms. Institutions are compelled to align their activities with the impact agenda, a task that is continually reshaped by new policy interventions. This dynamic creates both challenges and opportunities, necessitating a strategic and adaptive approach from higher education institutions to successfully manage and leverage research impact.

The chapter reveals that while the concept of innovation and the commercialisation of research are central to current UK policy, this focus brings about significant implications for universities. The integration of "research culture" and "rigour" within policy rhetoric, alongside the emphasis on market-driven innovation, illustrates the complexity of the frameworks within which institutions operate. Universities must balance these demands with maintaining ethical standards and fostering a research culture that values both academic integrity and societal relevance. The dual pressures of demonstrating economic benefits and adhering to rigorous ethical practices create a nuanced environment for institutions to navigate, and one which we argue they aren't always well-equipped to navigate in the best interests of impactful research.

Finally, the chapter highlights the pervasive influence of neoliberalism on the higher education sector, which manifests in various ways. From the pressure to generate income through innovation to the necessity of engaging in market-based strategies for sustainability, universities are entrenched in a competitive and market-oriented landscape. This neoliberal influence extends beyond the recruitment of students to the need to showcase impactful research to attract funding and support. The resultant institutional strategies often reflect a reactive approach to policy demands, emphasising the need for a more coherent and deeply integrated understanding of research impact within the academic and administrative fabric of universities. This chapter thus calls for a re-evaluation of how research impact is conceptualised and managed, advocating for a strategic alignment that genuinely embeds impact within the core mission of higher education institutions.

Notes

1 www.ukri.org/who-we-are/who-we-fund/#:~:text=We%20provide%20funding%20to%3A,(NGOs)%20and%20other%20institutions
2 www.fasttrackimpact.com/post/my-take-on-what-the-initial-decisions-on-ref-mean-for-impact [Accessed 30 June 2024].
3 www.icureprogramme.com/?utm_source=googleads&utm_medium=cpc&utm_campaign=icuresearch&gad_source=1&gclid=EAIaIQobChMIxYit-MyFhwMVyIxQBh1FCAIaEAAYASAAEgKAMfD_BwE [Accessed June 2024].
4 https://arcaccelerator.io/ [Accessed June 2024].
5 https://aspect.ac.uk/community/ [Accessed June 2024].
6 https://staffnet.cs.manchester.ac.uk/reso/refguidelines/REF-pathway-to-impact-guidance-nov2015.pdf [Accessed June 2024].
7 https://results2021.ref.ac.uk/profiles/institutions/10000975#:~:text=Introduction%20to%20the%20REF%20results,quality%20profile%20for%20each%20submission [Accessed June 2024].

References

Boeren, E. (2023). Investigating the relationship between research income and research excellence in education: Evidence from the REF2021-UoA23 data. *British Educational Research Journal*, *49*(6), 1312–1337.

British Academy. (2019). *Lessons from the history of British science policy.* British Academy.

Chandler, J., Housley, C., Rough, E., Hutton, G. (2021). *Research and development funding policy.* House of Commons.

Chen, R. (2007). The Politician's Fallacy and the Politician's Apology. https://devblogs.microsoft.com/oldnewthing/20070226-01/?p=27853 [Accessed June 2024].

Civic University Network. (n.d.). Capturing and Enhancing the Impact of the Civic University: Current Thinking, Issues and Challenges https://civicuniversitynetwork.co.uk/wp-content/uploads/2022/04/Capturing-and-Enhancing-the-Impact-of-the-Civic-University.pdf [Accessed 30 June 2024].

Daybell, J. (2019). The future of early modern studies: Embracing the impact and public engagement agenda. *The Sixteenth Century Journal*, *50*(1), 351–360.

DBIS. (2016). *Building on success and learning from experience: An independent review of the research excellence framework.* Department for Business, Energy and Industrial Strategy.

Dobson, J., & Ferrari, E. (Eds.). (2023). *Reframing the civic university: An agenda for impact.* Palgrave Macmillan.

Jarvis, S. (2023). How should universities understand their social impact?. In Dobson, J., & Ferrari, E. (Eds.), *Reframing the civic university: An agenda for impact* (pp. 41–61). Springer International Publishing.

Kelly, A. (2016). Funding in English universities and its relationship to the research excellence framework. *British Educational Research Journal*, *42*(4), 665–681.

Kelly, A. (2022). The end of the line? A quantitative analysis of two decades of competitive assessment and research funding in UK higher education. *Assessment & Evaluation in Higher Education*, *48*(5), 701–722.

Kline, S. J., & Rosenberg, N. (2010). An overview of innovation. Studies on science and the innovation process: Selected works of Nathan Rosenberg, 173–203.

Larner, W. (2003). Neoliberalism?. *Environment and Planning D: Society and Space, 21*(5), 509–512.

Maisuria, A., & Cole, M. (2017). The neoliberalization of higher education in England: An alternative is possible. *Policy Futures in Education, 15*(5), 602–619.

Perren, L., & Sapsed, J. (2013). Innovation as politics: The rise and reshaping of innovation in UK parliamentary discourse 1960–2005. *Research Policy, 42*(10), 1815–1828.

Pinar, M., & Horne, T. J. (2022). Assessing research excellence: evaluating the research excellence framework. *Research Evaluation, 31*(2), 173–187.

Research England. (2023). Research Excellence Framework 2028: Initial Decisions and Issues for Further Consultation. www.ref.ac.uk/publication/initial-decisions-on-ref-2028/ [Accessed June 2024].

REF. (2011). Decisions on Assessing Research Impact. https://2014.ref.ac.uk/pubs/2011-01/index.html

Smith, K. E., Bandola-Gill, J., Meer, N., Stewart, E., & Watermeyer, R. (2020). *The impact agenda: Controversies, consequences and challenges.* Policy Press.

Snowball, J., Tarentaal, D., & Sapsed, J. (2022). Innovation and diversity in the digital cultural and creative industries. In Woronkowicz, J. (Ed.), *Arts, entrepreneurship, and innovation* (pp. 187–215). Springer Nature Switzerland.

Stevenson, C., Grant, J., Szomszor, M., Ang, C., Kapoor, D., Gunashekar, S., & Guthrie, S. (2023). *Data enhancement and analysis of the REF 2021 Impact Case Studies.* RAND.

Thrift, N. (2022). *The pursuit of possibility: Redesigning research universities.* Policy Press.

Tuckerman, L., Nelles, J., Walsh, K., & Vorley, T. (2023). Sustainable innovation policy: Examining the discourse of UK innovation policy. *Environmental Science & Policy, 145*, 286–297.

UCU. (2009). Response to the Research Excellence Framework Second Consultation on the Assessment and Funding of Research. www.ucu.org.uk/media/3601/Response-to-the-REF-second-consultation-Dec-09/pdf/ucu_REFresponse_dec09.pdf

UKRI. (2022a). *Budget allocations for research and innovation.* UK Research and Innovation.

UKRI. (2022b). *UKRI strategy 2022–2027: Transforming tomorrow together.* UK Research and Innovation.

Watermeyer, R. (2019). *Competitive accountability in academic life: The struggle for social impact and public legitimacy.* Edward Elgar Publishing.

Williams, K., & Grant, J. (2018). A comparative review of how the policy and procedures to assess research impact evolved in Australia and the UK. *Research Evaluation, 27*(2), 93–105.

Wilsdon, J. (2020). A requiem for impact?. *Wonkhe*, [blog] https://wonkhe.com/blogs/a-requiem-for-impact/ [Accessed June 2024].

5 Moving Policy into Practice

Introduction

In the previous chapter we outlined the way in which impact as a public-policy-led agenda has created huge growth within universities, both in terms of funds allocated to pay for impact support as well as changes in the number of people involved in this agenda. In this chapter we will look at the discourse of impact around REF2021, as well as in the years that have followed in anticipation of a future REF (currently REF2029), as such a key driver of this activity in the United Kingdom. With two REF exercises having now happened (REF2014 and REF2021), and impact firmly within other elements of research funding, it seems that impact should be fully embedded as business as usual – for academics and for institutions. Among policymakers there is perhaps the idea that institutions shouldn't be particularly concerned about REF2029 because they now know what they're doing with impact, and there is certainly no rush to provide more guidance, even with the impact component of the assessment having been re-conceived as "Engagement and Impact".[1]

There is no doubt that there has been a change in the discourse around academia and in how the role of the academic is understood, but evidence from our research indicates that impact is still positioned quite awkwardly within the structures, attitudes, and practices of those who work within this space. Interviewees from a whole range of institutions and perspectives directly stated that they considered that impact was, at best, only embedded in a patchy way. This is quite in contrast to the portrayal for the REF2021 (which was submitted at the end of March 2021 – delayed due to the COVID-19 pandemic) in which institutions understand what impact is, how to do it, and how to write impact case studies, now it was the second time in which they'd had to submit them. Impact had been around for near enough 10 years, and with 7 years since the previous REF there had been the time in which to create the structures, roles, systems, and knowledge of impact to support it within their institutions. In particular, those in impact roles in universities have demonstrated in forums and networks a more coherent level of impact literacy, as is the term used developed by Bayley and Phipps (2019, 2023), that perhaps didn't exist for REF2014. These professionals understood the complexities of

DOI: 10.4324/9781003454991-5

the task at hand for the REF, and some also did for impact beyond the REF, and there were tools to help with this, such as bespoke online systems for tracking and storing impact narratives and evidence.

It's clear too, based on the amount of four-star impact achieved, that many institutions did very well for impact in REF2021. Overall, 87.2% of impact case studies submitted to the REF2021 were assessed as being of three- or four-star quality. This is higher than the volume of research (84%) and environment components (86.5%) assessed to be of this same three- or four-star quality (Research Excellence Framework 2022: 4). As the cross-panel report (REF 2022: 5) states:

> Across the exercise, panels continued to be impressed by the diversity and richness of the impacts described in case studies. There was clear evidence of the positive effect of HEI investment in supporting and enabling impact in the submitted case studies – as often evidenced in the environment statements.

Impact in REF has worked, then. Institutions apparently now support and enable impact properly, and the quality of impact in the case studies demonstrate that this has been effective. And very good this is too for institutions, where a four-star impact case study in REF2014 was calculated to be worth approximately £324,000 to an institution in quality-related (QR) funding over the course of the REF cycle (Reed and Kerridge 2017).

So the REF is a big thing for institutions. It seems that many institutions identified their most likely case studies at least two years before the submission deadline – and certainly some started doing this almost immediately after submission to REF2014 – and the mantra within the sector was to collect evidence when it was available, to record information about the impact being developed, and to train academic staff so that they had the skills to be able to deliver, identify, and articulate impact; to think about impact alongside the other elements of their role, and to actually make it happen rather than rely on serendipity as for REF2014. The emphasis for REF2021 was more about how to achieve four-star impact – the highest score available – because that is where institutions would achieve much greater income and be able to really promote their achievements publicly – rather than simply to have case studies to submit at all.

This doesn't really explain then the seemingly still difficult relationship between impact for the REF and impact as such though, which we will unpack further in this chapter. Although we have perhaps reached some kind of point of stasis with impact, our research has brought up a wide variety of tensions within the system and highlights how impact managers and teams have sought to address these. In the context of the argument we are making, about how impact should, at the point of assessment, be able to drop out of ongoing impact work, it is clear from our interviews, ethnography, and analysis of sector-based documents and discourse that this is simply not the case. There is still a huge amount of work and institutional pressure to *get* people to do impactful activities, let alone to try to support them all to do it well. It is perhaps the case that some institutions have a greater resource base to establish sophisticated administrative functions and skilled machinery around impact, but it seems the sector is still grappling with how to really do this. There is still a question as to whether impact, as legitimised by the REF, is enough to embed impact in institutions, or if further change is needed for this to happen.

The Relationship between Impact in REF and Impact in Institutions

The REF clearly remains a hugely dominant and influential feature of the discourse around impact in the United Kingdom. When discussing impact with professionals within the sector, it isn't long before REF impact case studies are mentioned. At the Association of Research Managers and Administrators (ARMA) conference meeting of the special interest group for impact professional services staff from across the United Kingdom, there was an opportunity to share with others, and decompress from, the trauma of writing impact case studies for REF2021, which some felt still lingered, over three years on from submission. An experience from impact professionals, which appears to mirror the "affective auditing", or emotional toil of potent auditing processes on the academic managers involved in these processes of evaluation (Watermeyer, Derrick and Borras Batalla 2022). Impact for the REF therefore appears to be seen as a sort of necessary evil; the thing that both constitutes much of the imperative for impact generation, while also being a sort of difficult, stressful, and high stakes game to be played.

Indeed, there is also the question of whether impact and impact in the REF are experienced as one and the same in the United Kingdom. Institutions, as drivers for impact and sponsors of impact evidence, arguably see it as the same thing. It seems to be particularly true that impact is understood by senior leaders and managers to be the part where you demonstrate the benefit of research to society and beyond, and the REF is the main outlet for this. This was particularly apparent from the Freedom of Information (FOI) return where the number of impact staff returned as such tended to be relatively low. When corroborated with interviews we noted that some interviewees referred to whole teams of staff organised under structures that were labelled as "impact" with far fewer indicated in the FOI return, for example one such institution reported having at least 50 members of staff in their impact structure, but indicated in their FOI return that they had zero impact staff. This suggests to us that many only count in their number of impact staff those who directly support REF-related impact activity.

Another way of looking at it would be to include all those involved in supporting the generation of impact too, with a plethora of knowledge exchange, policy, and commercialisation staff, not to mention staff on large projects, which may now also have impact staff costed in (Sutton 2020), as well as communications and events teams, among others. While these roles in number have perhaps not grown in quite the same way as those for impact managers and similar have, there is clear evidence from our interviews with impact managers that they have either been directly repurposed and reframed to support the impact agenda in institutions, or that this has been done in an ad hoc and less strategic way, with several impact managers in large Russell Group institutions raising the question of, for example, how they can make their commercialisation teams more interested in supporting impact. Institutional understanding of which kinds of research and innovation staff support impact, and which don't, appears to be quite a grey area, resulting in impact in one part of an institution sometimes trying to leverage support from other parts of institutions that aren't particularly engaged with the idea of impact. This might be less of an issue if impact staff were purely there to support the capture of evidence, but with many tasked with supporting the development of everything to do with impact, including the different pathways associated with impact, and the capacity and ability of academics to develop impact from research, this is then more problematic.

There is still a lot of myth surrounding impact in the REF though, which should probably be surprising for something that's essentially a bureaucratic exercise. This appears to be partly as a consequence of the process of assessment and the vagueness of how scores were allocated to each case study. Beyond the criteria for four-star impact as "outstanding impacts in terms of their reach and significance", there is no further guidance, and different disciplinary sub-panels responsible for assessing these case studies develop their own peer review culture for what is valued as four-star within the context of the "generic assessment criteria" as set out in the Guidance on Submissions document. This means that those working in institutions are tasked with unpicking how impact case studies were assessed based on half-available data (individual scores for case studies can be worked out in some cases but are not published) and trying to identify what it was about particular case studies that resulted in the score achieved (not all elements of any one case study may have achieved the score it ultimately achieved), and there have been plenty of researchers trying to analyse case studies to better understand this (such as Chowdhury, Koya and Philipson 2016; Battistin and Ovidi 2022; Phillips et al. 2020).

As well as this there is constant review and slow availability of guidelines for impact assessment, the last of which was identified by one interviewee as being particularly problematic for smaller, less well-resourced institutions without sophisticated systems and less readily available data. Impact, as we know it from REF2021, is defined as the "'reach and significance' of impacts on the economy, society, culture, public policy or services, health, the environment or quality of life that were underpinned by excellent research conducted in the submitted unit" (REF 2019: 7). There is a push within institutions to "start the process even though we don't know what the guidelines are yet", to draft case studies based on the previous REF rules and template, and to assess where these case studies are against the REF2021 criteria.

It's hard to underestimate the extent to which conversations within institutions and the wider impact community within the United Kingdom try to fill the void where there is no guidance on what the requirements for impact (now for REF2029 "engagement and impact") will be for the next REF. Senior leaders want to know so they can start to judge the output from their academics, and those within the community appear to want to know so that they can make sure they're doing the right things. There is still then, quite a bit of instability of the meaning of impact, or at least the perception of instability, and the concern that the institution isn't necessarily collecting the right information and data. Indeed, further myths abound as to what can be used as evidence of impact, or even submitted as impact itself. We've come across rumours, for example, about whether letters testifying about impact can still be used, and there is often still confusion about whether impacts on teaching in higher education "count". In sector-wide discussions many seem to still sort of be working to the old guidelines while there is also quite some discussion and frustration at not knowing what the new guidelines are, what is meant by rigour and how this will be assessed, and what the inclusion of "engagement" now means. Ultimately, we argue that if there wasn't such a short-term approach to impact by institutions though, many of these things currently seen as a problem would dissipate. Perhaps the difference between REF as a further twist of neoliberal managerialism versus impact that is said to allow spaces for progressive work towards social justice (MacDonald 2017: 697).

Also interesting is that, while the purpose of the REF is to "assess the quality of university outcomes to support dynamic and internationally competitive UK university research that would make a major contribution to economic prosperity, national well-being, and the dissemination of knowledge" (Audretsch et al. 2022: 398), there is very little sense of how this relates to the way in which impact is valued within it, in that it doesn't directly appear to assess this. Two-star impact is still, for example, indicated to be "considerable impacts in terms of their reach and significance", let alone three-star impact which is defined as "very considerable impacts in terms of their reach and significance". The idea of impact as being about demonstrating internationally competitive research, while the sector constantly strives for four-star impact because that's where significantly more funding comes in, seems amiss. Alongside, Budtz Pedersen and Hvidtfeldt (2023) who argue that impact assessment should also assess the micro impacts which are tangible and observable, this also calls into question the relatively small selection of impact examples institutions are required to submit, when, as we discuss later, there are many more impacts viewed by institutions as being "un-REFable". Bandola-Gill and Smith's (2022) research demonstrates how REF impact case studies, as a unit of impact, play a performative role in understanding what impact is, and we would argue that institutional pressure to submit only four-star impact compounds this, not helped by impact consultants who offer to show how to achieve four-star impact in one short training session.

A further challenge for institutions also appears to be the extent to which they know about the impact that is happening, particularly where the research took place longer ago. REF2014 and REF2021 allowed for impact from research that was up to 20 years old, but analysis shows that the average time from the start of research to the impact is 10 years, with panels C and D being on average 3 years less than panels A and B (Stevenson et al. 2023: vi), so the average for social sciences, arts, and humanities is less than 10 years from the start of the reported research. Our research indicated that institutional memory is potentially a significant issue within many places, with many institutions not even being able to hold the memory of the previous REF. Some institutions have brought back impact and research managers who had left previously (sometimes because the institution didn't seem very bothered about keeping them), at higher cost, upon realisation by the institution that they didn't have any recent knowledge of their own research and impact to be able to respond to the need for impact case studies, let alone knowledge of up to 20 years previously, and particularly where it may not have been obvious at the time that the research would develop to have an impact. In theory this should simply emerge out of the academic staff putting their hands up to say "I may have an impact case study", but this behaviour by institutions, as well as interviews with academics, indicate that this rarely happens.

In contrast to the external enactment of impact in assessment and audit sits what is becoming known as impact culture (Reed and Fazey 2021; Reed 2022). This is broadly about how impact is experienced within institutions, including the extent to which it is supported in a broad sense of the word, as a legitimate and freely undertaken part of research, and how collegiate impact is as an activity. Reed defines it as "communities of people with complementary purpose who have the capacity to use their research to benefit society" (Reed 2022: 18). In relatively normative discussions about impact, a supportive impact culture is seen as almost the antidote to a culture that's driven by assessment requirements. Drawing on Rickards et al.'s (2020) conceptualisation of research impact as ethos, Reed and Fazey (2021) define an impact culture framework, with four defined types of impact culture within institutions. They argue that a culture that gives academics freedom, and those academics then open themselves up to building relationships with stakeholders flowing from ethical research, provides the most likely context for tackling the significant challenges of the present. Their findings suggest that, in parallel with research culture being seen as overly competitive, as reported by over 4000 academics competitive in a Wellcome Trust survey (Moran et al. 2020), institutions need to look at how to prevent that in their own impact culture.

It is helpful then to consider the way in which impact professionals, as key supporters and promotors of impact within institutions, view their role in relation to institutional impact culture, as well as the ways in which they must attempt to create and navigate impact culture within their institution and the sector more broadly. In contrast to current understandings of impact culture within institutions, where institutions are often seen in opposition to researchers, we demonstrate later how messy the idea of the institution as constituted by people is, and how impact professionals can have the critical capability to not simply be part of the machinery of their employer, compliant subjects either acting without agency, or perhaps fully complicit in an unhelpful impact strategy that comes down from the top (Reed et al. 2022).

In terms of the messiness of impact "on the ground", Watermeyer and Tomlinson (2022) detail a survey about the experiences of REF2014 academics submitted to it as part of an impact case study. They conclude that this experience has left academics feeling dispossessed of their impact - impact which has generally not been generated as a consequence of the REF but because their research is more applied or disposed to affect society. Although we recognise echoes from our own research within the discussion, what is interesting here is the explanation that these negative attitudes to REF impact as being because of a lack of continuous investment in it is "emblematic of universities' approach to REF-impact as a tightly choreographed pop-up performance" and impact managers interested only in furthering their own careers. There is no doubt, as we will go on to discuss, that impact across institutions is patchy and, as we've discussed elsewhere, the resources made available to support impact driven by REF cycles, but we argue that the idea of this being tightly choreographed, particularly for REF2014 when impact was such a new concept, gives the sector far more credit for truly understanding impact than is due. Watermeyer and Tomlinson (2022) identify academics who weren't even aware of their impact being submitted to REF in an impact case study. Their conclusion is that this is due to the institution taking over, whereas we would argue that it's important to understand that there's almost a chaos of ownership, where in a case study that's co-written by a lead academic and various impact professionals, there may be assumptions from both as to who has made contact with the other researchers to discuss the case study. There are many ways in which we argue that this element of chaos and "making it up as you go along" continue within institutions, particularly where institutional memory has been lost due to the lack of continuous investment in impact by institutions.

How Do Impact Professionals Navigate Their Roles?

We interviewed a range of impact professionals who were both within institutions (from small and specialist, to broad based post-92s, to large Russell Group institutions) as well as consultants (who had all previously been impact professionals in institutions) and others involved in supporting academics to develop impact but who weren't based in specific institutions. Regardless of their views of impact, key themes that emerged relating to the way in which roles manifest are the importance to these professionals of creating their own approach to the role, and the positive effects of impact, as well as the continuing challenges, they experience and observe.

In particular, we identified those within the sector who were contributing to the development of ideas, discourse, and practice for how to support impact institutionally because we were keen to capture the innovation of the sector, and to identify the ways in which these thinking and engaged professionals add value to the work of researchers. Other aspects of our research, though, particularly for example interviews with academics, highlighted that this practice isn't universal (and where the typology of impact professionals in Watermeyer and Tomlinson (2022) may be closer to the truth), and that where these professionals play a much more functional role there was a sense in which it was purely doing the bidding of the institution, or perhaps even taking on a zombie role of proliferating mechanics of impact for the REF without providing any support for academics, or without, indeed, having any real understanding of impact. Asking academics to write case studies outside of the context of developing their understanding of impact, such as through a conversation about ongoing impactful research, or providing signposting to, for example, funding or additional services to enhance what they were doing, was particularly negatively received but still appears to happen within the sector. Another aspect highlighted was where information about the REF was circulated in an undigested form. Yes, bringing this information into institutions is seen as useful, but the inability for impact professionals to translate it into the institutional academic context was seen as unhelpful.

Although we are using the language of impact professionals, in part to encompass the broader range of roles than those more formally named as "impact managers" or "impact officers", sometimes with a combination of other activities included (such as engagement, or knowledge exchange, culture, or evaluation), interviewees, when talking about the purpose of their roles, emphasised how the support structure they were part of was crucial for the success of impact initiatives within institutions, at both the researcher and institutional level. Instead of just providing support for institutional submission to the REF, they discussed how their dedicated impact roles mean that academics have access to guidance and support for impact activities to develop their impact literacy, but that they also support impact through a whole range of other activities, such as supporting academics to coordinate and plan their impact activities, helping academics understand the role of language and language barriers in working collaboratively, and supporting connections to organisations beyond the institution (which typically would be thought of as the public engagement or knowledge exchange space). This key capacity building role is seen by these professionals as fundamental to the sector's ability to "do" impact more fully and comprehensively, and indeed this relatively comprehensive approach was seen as fundamental to the role, with impact professionals seeking to work with any or everyone with the potential for impact from their research. One interviewee who had been in an impact management role for a full REF cycle explained how there have been noticeable differences in that time, particularly in disciplines where impact is often seen as harder to achieve, such as physics or modern languages, and that this has been achieved by the persistent presence and engagement of impact managers and other impact professionals in their disciplinary spaces. They explained that this was about raising the profile of impact, demonstrating impact to be a legitimate activity, and supporting basic knowledge of it.

Interviewees noted that for academics to develop the skills and expertise needed for impact work, ongoing training and professional development is required. This often takes place through workshops and seminars, although the bespoke nature of impact pathways means that researchers are often seeking and responding better to one-to-one support, with the term "coaching for impact" having been one such area that has been scoped out and identified as best practice within the impact community (as presented by a range of professionals from different institutions at ARMA 2024 in a session on "Coaching for Impact/Coaching for Research Support").[2] Additionally, others noted the value of platforms for knowledge sharing – particularly sector-wide platforms that are able to bring a sense of, what has been suggested by the ESRC-funded ARC Accelerator as "cultural currency" – so the use of these by impact professionals when working with academics can also bring a sense of legitimacy to impact and related activities. Where it came to sharing impact and impact techniques between institutions, there was a real understanding from impact professionals in these institutions that they had to have the authorisation from their institution to be able to do this, even though they almost entirely indicated that they didn't think it made any difference to the competitive outcome of impact. As one interviewee put it, the academic still has to actually do the work themselves to create the impact. Professionals were, in fact, keen to share between institutions and find out what others were doing in order to develop and innovate.

Competition between institutions, and the pressure from Deans and other senior leaders to do better because the impact ranking is currently lower than the output ranking, and the unquestionable understanding that particular departments must retain their top position for impact in future REFs was viewed as inevitable in the current context of impact assessment. There was though a feeling from professionals that their institutional leaders simply don't understand how messy and unpredictable impact can be. As one interviewee said, impact fails all the time. Not because it's not good work, or because there aren't the right stakeholders, or because it's not supported, but simply because of factors beyond the control of anyone involved. Big pieces of policy work, for example, carried out in conjunction with governmental departments over a number of years, frequently go unimplemented. Impact professionals also have a clear understanding that impact is still heavily reliant on there being academics with impact as core to their practice established over many years. One Russell Group institution puts their success with impact simply down to their fortune in having academics who do very impactful work. They recognise that as professionals they have no say in who is appointed to these roles, and that decisions about appointments are largely not taken on the quality of an academic's impact.

Professionals also identified other factors which limit how they're able to influence academic impact practices, with factors such as the integration of impact into strategic planning, resource allocation, workload planning, and promotion all factors that helped determine the ability for them to fulfil their role, but that were ultimately beyond their control. There were some examples of how institutions had given greater priority to impact in recent years, such as through the establishment of impact structures that also included a wider range of professional services staff supporting a broad range of impact pathways (e.g. policy, commercialisation, and public engagement), with the idea that taking a more comprehensive approach helps embed impact into the institution's culture and helps foster a more sustainable and joined-up approach to environment. The ability for academics to demonstrate the value of their impact through career frameworks (alongside other academic activities) was also discussed as being important but was beyond their control, and where it was embedded in these frameworks, the extent to which this was, in reality, a fully established and legitimate activity for academics was mostly questioned. As one interviewee observed, academics still probably prioritise academic publications and grant income over everything else, often meaning the perception is that there is little or no time for impact.

Many participants reported finding their way into impact roles without a significant career plan for this, with very little sense of there being a traditional route into impact management. Some of those we interviewed moved into impact roles from other areas related to research management in higher education, such as public engagement or research management itself, while others moved into impact management roles from previous academic positions. There was a clear sense from some of these interviewees that impact had become more firmly established within the sector than some other roles and therefore there was a greater opportunity to establish something more significant than they had been able to in other roles. In particular, those from public engagement backgrounds highlighted that the initial enthusiasm about public engagement as had been funded through, for example the 2008 Beacons for Public Engagement[3] and the 2012 Catalysts for Public Engagement with Research,[4] and there was a sense in which public engagement as a function within institutions had been defunded in favour of impact. As public engagement professionals though, they then saw the opportunity for public engagement with research as now existing within the impact agenda. At the same time, those from academic backgrounds often also came from a very applied and engaged position within academia and, despite quite a number of years on academic contracts, found it was not possible to obtain a permanent academic contract undertaking the kind of methodologically engaged research with communities that public engagement makes the case for. In one case, an interviewee had continuous employment in academics posts for a number of years, but was only on short-term contracts until their appointment to a permanent professional services post within an institution where they are able to co-lead a research unit as an impact manager. Likewise, there are other examples of those in non-traditional hybrid roles, and those with backgrounds in, for example, evaluation and knowledge exchange.

Apparent then is the vast range of skills that our interviewees bring to their institutions within the medium of impact. It is a catchall for a significant number of skills and is perceived to have perhaps overtaken other related areas for prominence and institutional support. So the journey into impact roles for these people is fundamentally facilitated by institutional changes and external policy, notably the REF, and has brought about a structural change that has provided pathways for individuals to engage in impact work more formally. And while the route into impact roles may start as a move away from what we might think of as more traditional academic or professional paths, it still seems to be in a discursive space as to who and what an impact manager role comprises. The REF may be the driving force for the creation of the majority of these roles, but it certainly doesn't seem to be the motivation for interviewees in taking up these posts. Instead, they saw it as an opportunity to extend and develop the work they'd been undertaking in these perhaps more valued spaces of non-audit. One senior manager within a large and highly ranked institution described how they wouldn't expect staff to have a research degree, but described looking for those with intellectual curiosity and sharpness of mind; perhaps similar characteristics to those seen in researchers, and it is also notable that quite a number of those in impact roles, and particularly those we interviewed, do have higher degrees or PhDs.

The reason for working in impact roles was often cited as being the value they see in doing the work; as there being a sense in which their skills could be used and are relevant for the benefit of society and beyond. Rather than simply seeing an impact role advertised and seeing it as a way of getting paid employment, they often made the move after exposure to concepts of impact through academic conferences, workshops, or conversations with colleagues who were already engaged in impact-related activities. It was seen as a way of doing a job which was in keeping with not only their general skills but also, for some, their very specific skills around how to engage, how to evaluate and use evaluation in a developmental way, and an openness to innovation and dialogue to support the fulfilment of the value of research to society. These varied pathways into impact roles start to demonstrate the extent to which impact management is populated by those who aren't just information or knowledge managers or auditors. Rather it is seen as the new name, or the new location, of resource that encompasses a wide variety of skills and knowledges that are intrinsically valued rather than valued for their instrumental value to an institution. There is no one place they come from and, as such, embody a plurality of skills and experience and aren't a homogenous group that "manages" impact.

In parallel with the position of impact within academic careers, there was a concern that the value of their role within frameworks that are recognised sector-wide is relatively precarious, particularly where they outlined the relationship between their academic identity alongside their identity as an impact professional. Several had developed their role within their own institution and created quite bespoke roles with both impact support and academic elements which gave them the opportunity to practice both of these aspects of their identity. They did this, for example, by developing with their institution quite unusual "hybrid" roles, or through promotion along an academic pathway while also retaining their key function and identify as the institutional impact manager or similar. This was seen to be quite an experimentation, and only possible because of the personal support from their own managers. They also noted how fundamental the unconventional hybrid nature of their role was in them supporting impact well, but also indicated how they felt it was unrecognised as a model across the sector. These interviewees expressed concern about the extent to which they would be unemployable in other institutions should they decide to seek a role elsewhere.

Recognition and career progression within impact roles vary significantly across institutions, and interviewees who highlighted the unusual nature of their role indicated that they expected more traditional academic institutions to be less open to this degree of innovation and identity bending. Although the idea of career frameworks which include impact are common within academia, and there is a sector-wide understanding that impact should perhaps be embedded within these progression frameworks if impact is to fully be embedded, it seems that what is traditionally thought of as professional services may also not incorporate impact as a pathway particularly well. The combination of academic and research management skills (at the very least) sits quite uncomfortably with the traditional split between the producers of research and impact (academics) and those who structure the support around these people (professional services), when these impact professionals may well be researching and publishing themselves.

Institutional impact leadership was a particularly interesting theme to emerge from the interviews with impact professionals. Several of these interviewees talked around the idea of leadership without directly referring to it, but there is a sense in which there is not only a leadership gap for impact within senior leaders in institutions, but actually that these leaders aren't necessarily best-placed to provide impact leadership. One interviewee directly addressed this issue, and discussed that the sort of people qualities and skills required to create impact does not sit easily with many of the sort of people who currently lead institutions who don't especially value the sort of qualities that they suspect impact professionals often bring to their roles. The interviewee highlighted these individual qualities and skills such as being able to see all the deep richness of detail, but also being able to take it to the level of the bigger picture and being the "ultimate connector". They also suggested (though indicated this was not based on evidence, more anecdote) that impact professionals perhaps also tend to be more nurturing of people - something that plays out in their everyday activities working with researchers - and that this is a different but fundamental foundation for impact leadership, and one that institutions don't acknowledge or value enough. Having some of these personality traits in impact roles then seems to mean that impact professionals can find themselves trying to bring about a revolution and culture change on their own, without the support of institutions. On the converse we might also observe that where impact professionals are perhaps seen to embody "zombie"-type roles, the lack of demonstration of nurturing characteristics, and their engagment at a more superficial level with both research and the broader picture (alongside reduced knowledge of research impact), means there is a less effective version of impact leadership and support.

Relatedly, the question of gender dynamics and power inequalities was discussed, both in terms of women being disproportionately represented in impact professional roles (although again, this was observation rather than evidence), and how this may be because of the gendering of types of work and that impact support often requires care, support, and engagement skills. Impact as being about "emotional work" is also something that emerged as a perception with REF2014 panellists (Watermeyer and Chubb 2019), and empathy is also highlighted as fundamental to impact by Reed (2022). While in a professional support role, one interviewee was involved in writing a paper from a particularly engaged research project relating to policy (where those doing the engagement work were almost exclusively female) was immediately rejected for publication on the basis that it was not seen to be a legitimate academic research methodology; it didn't fit the world-view of the journal editors, and they entirely missed the point of it. With regard to the discussion about leadership, this may demonstrate the persistence of a patriarchal, hierarchical culture within higher education institutions, which continues to privilege competition over collaboration.

Gender was also highlighted by interviewees as significantly influencing the perception and value of impact work within academic institutions and academia more widely. This view not only corroborates "the downstream gendered inequities in HE such as through an impact agenda" (Chubb and Derrick 2020) but also develops this argument through an understanding of how these inequities may be further amplified through the positioning of those with impact leadership skills in professional services roles that are traditionally seen as administrative rather than leadership. The inclusion of more of these diverse views and experiences of those who work as impact professionals – addressing equality and diversity within leadership structures and decision-making processes – may support the embedding of impact within academia. Although none of the interviewees referred to direct experiences of been viewed as "less than" or as failed academics, in the way that Watermeyer discusses with public engagement professionals as part of a prestige economy (Watermeyer and Rowe 2022) there is still evidence to suggest they are positioned in this way in institutional structures.

Perhaps as a consequence of the deep thinking and doing work that many of our interviewees consider core to their role and identity, the ethical landscape of the impact as viewed through the lens of it as an "industry" emerged as a major concern for many interviewees. There is a firm recognition that, like any sector, the impact industry has the potential to exploit vulnerable populations when such emphasis is placed on researchers to demonstrate impact from their work. With this being such an imperative, there is significant concern that the power of highly educated individuals within large and comparatively wealthy institutions can effectively arrive into places perceived to have fewer options and opportunities for employment, education, and significantly less monetary wealth and that the government earmarked for "levelling up". There is a concern both that these are viewed as easier clusters of people to access and establish change with, and particularly that evidence of change will be easier to collect (or unethically coerce) from them for impact assessment. This is not necessarily to say that it happens, but does demonstrate the extent to which impact professionals are engaged beyond assessment and audit, with a commitment to impact being established in ethical ways, and driven by genuine intentions to create positive change, rather than write a simplified narrative for case studies which emphasises the positive impacts, while neglecting to include elements of harm (termed by Derrick et al. (2018) as "Grimpact"). For example, several interviewees discussed how they were seeking, through their own research, to develop deep evaluation practices based on participatory methodologies and creative practice, to establish new versions of best practice which were not only able to better understand and articulate the value of more engaged forms of research but that also understand the role of evaluation as part of the way in which participants experience research itself. Evaluation, therefore, has a role to play in ethical research and impact practices.

Those we interviewed often emphasised that the real value they found within their role was their ability to help researchers understand what impact is and how to develop an approach to embedding impact within the specific context of their research. Although REF impact case studies came up in all interviews this was recognised as a process, and achieving four-star impact case studies within their own institutions was viewed as strategically important but not referred to as the main interest to them in their roles. Indeed, many emphasised the value of the smaller, currently less REFable, perhaps more community-based version of impact they have supported, and the significant value to the aims of the researchers and to those outside the institution of doing this. They discussed the problematic nature of superficial metrics that may inflate impact without genuinely contributing to societal good. Instead, they discussed focusing on substantive, long-term benefits that their work can deliver to communities and individuals.

Despite the deep ethical thinking of impact professionals, ethical challenges were seen to persist, primarily within the institutional context and the in actions seen as "the institution", beyond their control. Participants highlighted issues such as the "airlifting" of high-profile speakers who contribute little actual value, creating the illusion of engagement rather than real, impactful change. They emphasised the need for more systematic and authentic engagement practices that prioritise meaningful contributions over appearances. This ongoing struggle underscores the complexity of maintaining ethical practice in a field driven by both altruistic goals and external pressures to demonstrate impact.

In addition to professional identity and leadership, the notion of impact professionals needing to constantly develop innovative practice in order to support impact was discussed in relation to impact pathways. Despite the greater availability of funding for impact in some institutions (though some highlighted challenges around this with their lack of funds such as HEIF and Impact Acceleration Accounts which are only in more plentiful supply for some institutions), the ability for impact to be sustainable was seen as a significant challenge. Some institutions take a "whole pathway" approach to trying to identify what might be most suitable for any one academic, with those specialising in a whole range of approaches coming together to discuss how the different pathways might be able to support impact. The level of organisation and resource required for this though, not only in the initial meeting but also for the ongoing support, is likely beyond that most universities would agree to commit to. Commercialisation in the SHAPE disciplines was discussed as one of the newer innovations, and indeed greater commercialisation leading to impact was also flagged by panel A in their review of REF2021 impact, but is still relatively small in scale, with supportive innovators in professional services role being key to how this is being communicated to academics and developed.

Alongside this, and in reflecting on their own career pathways too, impact professionals are acutely aware of the difficulties for some academics in transitioning from more traditional academic roles to those which are focused on delivering impact from their research. One significant issue that was highlighted is imposter syndrome, where individuals doubt their abilities and feel they do not belong in their changed roles, and professionals often see this exacerbated by the differing skillsets required for impact work compared with traditional academic research. This is seen across a range of impact-related activity, but particularly for commercialisation, where working with academics isn't just about providing them with the skills to commercialise their research, but is also about having to bring about a shift in mind-set. Although Watermeyer and Tomlinson (2022) would likely challenge that this makes impact professionals part of the psychological amnesty of the creation of competitive accountability within institutions, and therefore all part of the agenda, impact professionals (some of whom are still cognisant of the balance between commercialisation and research) see this as part of the way in which academics can expand and develop their research, while also creating social benefit.

The final key theme that emerged from interviews with impact professionals was in relation to the need for robust evaluation methods and the gathering of evidence, which are seen as critical components of impact work. Interviewees emphasised the importance of systematic evaluation to demonstrate the effectiveness and reach of impact activities. There was a clear question about the extent to which there is yet a sector-wide understanding about how to establish evidence to demonstrate the outcomes of impact projects and the role that evaluation can play within this. Interviewees noted that this is essential for securing funding, gaining recognition, and improving future efforts, but that there is still a significant lack of engagement from the academic community in implementing these approaches. It was noted that, although flawed, other sectors must evaluate far more fully out of necessity for funding, and that it is baffling that there is such poor knowledge of this within academia given the role of academic researchers in creating and using research data. It was felt that academics are not trained to have "a proper moment of reflection", that they tend to be future focused and therefore often do not have a clear understanding, or even interest, in what has happened as a consequence of their past work. It was thought this was possibly as a result of systems and structures that do not require this, or provide time, funding, and opportunity for it, and was another example of where there is now a gap between what academics want to do and achieve (change and benefit from their research beyond academia) and these systems and structures.

The need for innovation in gathering evidence of impact was also seen as essential to embedding impact more fully, particularly as evidence gathering can be complex and time-consuming. Interviewees highlighted the need for creative and inventive methods to collect and present evidence, to supplement approaches such as written testimonials and quantitative metrics with more innovative techniques like video testimonials, participatory evaluations, and using interactive platforms to provide richer, more nuanced insights into the impact of a project. The continuing lack of impact literacy among academics and institutions was felt to be quite a barrier to achieving this.

Conclusion

In examining the role that impact professionals can play in constructing impact support for researchers, we suggest that many spend as much time navigating the institutional maze in order to do what they see as one of their main roles, which is to embed impact in a holistic way, as researchers do. They drive a huge amount of innovation within the sector, and many provide a significant amount of leadership for impact, which is often unrecognised by their own institution. We suggest that there is still a significant amount of research to be done to fully understand how impact professionals can be enabled to establish these more holistic impact cultures, and to have these more fully legitimised by their institutions. The key role professionals play in interpreting policy in the context of being in a minoritised role has started to be examined in some contexts (such as McKay and Robson (2023) in relation to the Teaching Excellence Framework), but professional service roles aren't homogenous and there's more to be understood in relation to impact professional practice.

Rather than being institutional subjects, impact professionals often have a significant and critical contextual understanding of the neoliberal agenda around impact, and will act in ways to attempt to mitigate the effects of this, but they tend to do this as a consequence of their own identity rather than necessarily because of an institutional driver to do this. They may see this as the only real way to embed impact legitimately within the sector, but again, this may not be fully acknowledged within the institutional structures and processes.

These roles may be staffed by individuals who take on many of the same fundamental ethical questions about impact in academia as those who carry out the research, and they attempt to find value and benefit beyond audit and assessment. However, evidence suggests that this may be by accident of appointment to the role rather than something that was initially designed into the role, and so the sector should consider how they can design these roles in such a way as to access this level of ethical engagement, innovative thinking, and leadership for impact through roles that have traditionally been in the professional services domain. Alongside this though, institutions may do well to better or further integrate these roles into their leadership structures to support a more embedded and holistic version of impact that is being more formally driven for reasons beyond assessment and targets. Rather than being seen as part of the problem, which the literature often seems to, these professionals may well be actively working to lead a more radical and empathetic form of impact.

Notes

1 www.jisc.ac.uk/future-research-assessment-programme/initial-decisions
2 https://abbey.eventsair.com/AbbeyEventApp/arma-2024/agenda/Agenda/AgendaItemDetail?id=4ad26b4d-55d8-4243-adee-7f85557d10f9 [Accessed 30 June 2024].
3 www.publicengagement.ac.uk/resources/report/beacons-public-engagement [Accessed 30 June 2024].
4 www.publicengagement.ac.uk/resources/report/beacons-public-engagement [Accessed 30 June 2024].

References

Audretsch, D. B., Belitski, M., Guerrero, M., & Siegel, D. S. (2022). Assessing the impact of the UK's Research Excellence Framework on the relationship between university scholarly output and education and regional economic growth. *Academy of Management Learning & Education*, *21*(3), 394–421.

Bandola-Gill, J., & Smith, K. E. (2022). Governing by narratives: REF impact case studies and restrictive storytelling in performance measurement. *Studies in Higher Education*, *47*(9), 1857–1871.

Battistin, E., & Ovidi, M. (2022). Rising stars: Expert reviews and reputational yardsticks in the research excellence framework. *Economica*, *89*(356), 830–848.

Bayley, J., & Phipps, D. (2019). Building the concept of research impact literacy. *Evidence & Policy*, *15*(4), 597–606.

Bayley, J., & Phipps, D. (2023). Extending the concept of research impact literacy: levels of literacy, institutional role and ethical considerations. *Emerald Open Research*, *1*(3), 1–15.

Budtz Pedersen, D., & Hvidtfeldt, R. (2023). The missing links of research impact. *Research Evaluation*, *33*, 1–8.

Chowdhury, G., Koya, K., & Philipson, P. (2016). Measuring the impact of research: Lessons from the UK's Research Excellence Framework 2014. *PloS One*, *11*(6), e0156978.

Chubb, J., & Derrick, G. E. (2020). The impact a-gender: gendered orientations towards research Impact and its evaluation. *Palgrave Communications*, *6*(1), 1–11.

Derrick, G. E., Faria, R., Benneworth, P., Pedersen, D. B., & Sivertsen, G. (2018). "Towards Characterising Negative Impact: Introducing Grimpact". Proceedings of the 23rd international conference on science and technology indicators: Science, technology and innovation indicators in transition (Vol. 2018-09-11, pp. 1199–1213). Leiden University, CWTS.

MacDonald, R. (2017). "Impact", research and slaying Zombies: The pressures and possibilities of the REF. *International Journal of Sociology and Social Policy*, *37*(11-12), 696–710.

McKay, F., & Robson, J. (2023). 'Structured agency', normalising power, and third space workers: higher education professional services staff as regulatory policy actors. *Journal of Further and Higher Education*, *47*(5), 633–646.

Moran, H., Karlin, L., Lauchlan, E., Rappaport, S. J., Bleasdale, B., Wild, L., et al. (2020). Understanding research culture: What researchers think about the culture they work in. *Wellcome Open Res.*, *5*(201), 1–32.

Phillips, P.A., Page, S.J., & Sebu, J. (2020). Achieving research impact in tourism: Modelling and evaluating outcomes from the UKs Research Excellence Framework. *Tourism Management*, *78*, article 104072.

Reed, M. (2022). *Impact Culture*. Fast Track Impact.

Reed, M., & Fazey, I. (2021). Impact culture: Transforming how universities tackle twenty first century challenges. *Frontiers in Sustainability*, *2*, 662296.

Reed, M., Gent, S., Seballos, F., Glass, J., Hansda, R., & Fischer-Møller, M. (2022). How can impact strategies be developed that better support universities to address twenty-first-century challenges?. *Research for All*, *6*(1), 1–33.

Reed, M., & Kerridge, S. (2017). How much was an impact case study worth in the UK Research Excellence Framework?. *Fast Track Impact*, *1*, 47–49.

REF (2019). Guidance on Submissions (2019/1). REF2021. https://2021.ref.ac.uk/publications-and-reports/guidance-on-submissions-201901/index.html [Accessed June 2024].

REF (2022). Summary report across the four main panels. REF2021. Available online at: https://2021.ref.ac.uk/media/1909/cross-main-panel-report-final-updated-september-2022.pdf

Rickards, L., Steele, W., Kokshagina, O., & Morales, O. (2020). Research Impact as Ethos. RMIT University, Melbourne, VIC. Available online at: https://cur.org.au/cms/wp-content/uploads/2020/09/rickards-et-al-2020-research-impact-as-ethos.pdf

Stevenson, C., Grant, J., Szomszor, M., Ang, C., Kapoor, D., Gunashekar, S., & Guthrie, S. (2023). *Data enhancement and analysis of the REF 2021 Impact Case Studies*. RAND.

Sutton, E. (2020). The increasing significance of impact within the Research Excellence Framework (REF). *Radiography*, *26*, S17–S19.

Watermeyer, R., & Chubb, J. (2019). Evaluating 'impact' in the UK's Research Excellence Framework (REF): Liminality, looseness and new modalities of scholarly distinction. *Studies in Higher Education*, *44*(9), 1554–1566.

Watermeyer, R., Derrick, G. E., & Borras Batalla, M. (2022). Affective auditing: The emotional weight of the research excellence framework. *Research Evaluation*, *31*(4), 498–506.

Watermeyer, R., & Rowe, G. (2022). Public engagement professionals in a prestige economy: Ghosts in the machine. *Studies in Higher Education*, *47*(7), 1297–1310.

Watermeyer, R., & Tomlinson, M. (2022). Competitive accountability and the dispossession of academic identity: Haunted by an impact phantom. *Educational Philosophy and Theory*, *54*(1), 92–103.

6 Good Research Can Be Impactful Research

Introduction

"If you do good research and have good impact, research assessment is simply something that comes around, not something you should focus on."

The above quote is taken from a former colleague of ours, a now retired former DVC and a person who remains active in his research discipline because of his interest and passion in his field. Unusually (as indicated by other academics we have interviewed) for a senior leader, this one someone who went out of his way to inspire and support others while still being a research leader in his own discipline.

He was fond of saying this in meetings about research policy and Research Excellence Framework (REF), particularly in those meetings where research and academic managers would be planning for "mock REFs" and "readiness exercises" with many reviews about the current state of outputs, whether academics were submitting their case study "evidence" to central systems and feedback from external consultants on their views on where academics needed to be told to do more work. His view was always that the most important thing about an effective research assessment return was creating a research culture where good research could be done.

In unpicking this quote, it is perhaps is an excellent illustration of the central premise of this book – that if impact is not part of the research process, and if academics (or at least some academics) do not see its value and how it applies to their values, beliefs, and identity, it will remain poorly understood and done as a bolt on to research activity that we are doing "because we have to" or as a result of funding demands, rather than because we want to.

If one is active in their field, regularly publishing and engaging with stakeholders (should the research work be appropriate to be transferred beyond academia), who then take up the research findings and act upon them, research assessment exercises are things that come around and you can contribute to them.

Far more problematic, we would propose, are those who view research evaluation instead as an institutional goal, rather than part of systems and policies that surround academic practice. We have frequently heard from

DOI: 10.4324/9781003454991-6

colleagues and peers about how they now "have my REF papers sorted" or research managers saying "We've got enough now", as if this is the end goal of any academic activity. Certainly, a reflection on management excess within academic culture, if we are to move impact, which is still perceived by many as something poorly understood that we have to do regardless, to something academics do as a fundamental part of their research practice, impact culture needs to be wrestled away from senior leaders and metrics-obsessed managers back to the often neglected grass roots of the academy.

In considering where the sector is at with research impact in the previous chapter, this chapter argues for a deeper appreciation of research impact as a fundamental part of the research process, and the need to decouple impact as an add-on for research assessment. The chapter also moves the narrative from why we do impactful research to how we might do it better. It begins by arguing if we have, as a result of metric-driven research policy, lost our understanding of a fundamental importance of research to universities in terms of knowledge generation – dissemination that will change and improve society and align with the public good – we will never realise the potential for impact to be part of the academic identity.

Moving from the Shallow End

This chapter explores the attitudes of academics we interviewed as part of this project. In conducting these interviews with (as defined in Chapter 1) a group of academics from different disciplines, and different career stages, but all of whom had already demonstrated significant impact, in terms of both REF2014/2021 and also among the stakeholders that comprise the academic's disciplinary ecosystems, we wanted to explore why impact was important to them and what support they had received by peers and institutions as their careers progressed. In doing so, we hoped to unpick why impact is still poorly understood and managed, yet, conversely, these academics were doing it regardless of this, as they saw it as a fundamental part of their academic identities. This, in turn, would help us better understand how to move from a shallow impact culture carried out because academics felt they had to, to deep impact.

However, prior to exploring these discussions in detail, it is worth returning to a point we made in Chapter 1 – at the start of this project we had some concerns (and perhaps a little imposter syndrome) that maybe the problem was not the sector, it was us. We had become so obsessed with doing impact "correctly" and being ideologically pure regarding our impact motivations that anyone who didn't buy into the vision was wrong. Maybe our conclusion that a top-down, policy-driven approach to impact is the reason institutions still poorly understand it was incorrect and maybe the sector was awash with institutions where impact was treated as importantly as outputs, and it was simply an issue that we had been working in and with the wrong institutions!

It became very clear, very quickly, that the vast majority of our interviewees experienced similar frustrations and had experienced attempts to control or take over their research ecosystems. In a lot of cases, they had careers hampered by line management and leaders who could not see the value in what they were doing or, in the worst cases, actively tried to demotivate individuals and block their progress.

We would argue that, in many cases, good research is intrinsically impactful research. That is not to say that all academics should be doing it, but they should have the opportunity to do so as part of their research culture if it fits with their ideology, and they should expect support and recognition from their institutions if they do. Even the most theoretical of research has, at its core, a need to further knowledge for the benefit of society and to transfer that knowledge from academia to society (or, it should). In this chapter we use our interviews and personal reflections to explore what impact is, but also what it is not. And we also explore why people reach these misunderstanding, and the role of research stakeholders in cultivating this poor knowledge.

As we are fond of doing in this text, we will start with an anecdote, this one from Andy's early career.

One on my earliest exposures to the world of impact (even though it would never have been called that back then) was in 2006, just as I was starting to get involved in the tech policy world. A report with British Telecom (Lacohée, Phippen and Furnell 2006) had received considerable attention in policy circles, its focus being on the trade-offs between convenience, trust, and security expressed by end-users making use of online systems. The work was of interest to industry but also policymakers who were increasingly looking at the responsibilities of technology providers in ensuring the safe and secure deployment of IT systems. As a result, I received an invitation to attend a policy event in London that explored privacy and security. At the event a senior, very inspiring academic was presenting, and unpicked the recently announced NHS Connecting for Health programme (McGrath et al. 2008) which was attracting a great deal of press attention for both its ambitions to centralise healthcare records (called the National Programme for IT) in the United Kingdom and provide access to medical practitioners in a manner never attempted before, and its costs, which were considerable.

During the presentation, the academic was damning about the intentions of the system, and talked about the wealth of literature that highlighted the security issues associated with large-scale federated systems and the potential for massive data breaches that would significantly impact upon patient safety and privacy. In his view, it seemed a ridiculous proposal because the literature was so clear that it would fail. He was evidence-led and passionate in his delivery, predicted that the proposed new system would be a colossal waste of money and explained why. He expressed frustration that, given the academic work in this area, the government would be so willing to spend so much public money on a system that, in his view, would clearly never achieve its aims.

Someone from the Department of Health spoke next at the event and made the statement:

> "We don't have the time to read academic literature."

And subsequently discussed the various mechanisms for academics to provide input into policy choices beyond simply writing outputs and expecting the wider population to read them. These included select committees, consultations, and briefing documents.[1] At no point did they answer any of the academic's questions about why they could so recklessly propose something that was clearly not going to work (in their view), just that if academics wanted to influence policy, they needed to come to them.

This simple vignette was my first exposure to the tension between the expectations of academics that the research they publish has public value and should therefore be read by those making policy decisions, and policymakers who viewed academic literature cumbersome and time-consuming.

As an aside, Connecting for Health ceased to exist in March 2013, having spent an estimated £20 billion. At the time of writing (April 2024) the UK government has just announced a new large IT project to bring £36 billion of cost saving to the public sector (the majority being in the NHS).[2]

This more recent example occurs against the backdrop of research impact being part of research assessment since, approximately, 2011 in the United Kingdom. So, after all this time, the challenges of influencing policy remain, which would suggest little progress has been made in bringing the academic community closer to the stakeholders they wish to influence. Should we conclude, therefore, that impact is not worth bothering with? Certainly not, as we will see through this chapter, there is much extremely effective impact going on that is making real change to stakeholders, whether they be industry, policymakers, or the public at large. However, we might be right to ask "Why isn't this policy drive for great research impact working?"

Impact as a Management Function

In this chapter, where we are trying to adopt an evidence-informed exploration around why impact has not become part of academic identity, we would argue that there are fundamental challenges when impact is viewed as something that has first been ignored by university leaders, and then viewed as a policy initiative that needs to be managed due to the eventual financial incentives for "doing" impact in their institutions. If, as we argue, impact is still often viewed as a management-driven obligation, we can begin to unpick why impact is so poorly understood in institutions.

By way of illustration given further are a number of what we might refer to as "impact myths", all of which we have observed when engaging with both academic and management staff across the sector. There are also a number recognised by peers we have interviewed. We present these as a starting point to both illustrate our concerns about the understanding of impact in the sector but also to start out exploration of how to move on from here.

- **Everyone should do impact:** Some senior managers seemed to react to the announcements about impact in the REF by declaring that everyone with responsibility for research should now do impact as well. Clearly this is not the case and makes little sense given that some forms of research lend themselves to impact far more so than others.
- **We only want one or two people doing impact:** Conversely, some managers will, instead, decide that, depending on the size of a submission for a particular unit of assessment, only a couple of people need to "do" impact, because that is all that is needed for research assessment purposes, reflecting more about management culture than an appreciation of why academics do research.
- **I can do impact instead of publishing:** Some academics have expressed a view that they wish to learn more about impact because they think they would be better at that than writing articles. This is perhaps a reflection on the "published or be damned" view of toxic academia (see later), but unfortunately, we have to dissuade academics of that view, given that without underpinning research, there is no research impact.
- **You don't have an output you've had impact from:** A curious statement heard from a number of interview subjects and also from various impact training/REF events we have been in attendance at. A view that impact has to be linear, linking a paper to an "impact event", whereas a lot of impact will arise from a critical mass of research, rather than writing a single paper then "doing some impact" with it.
- **Doing a research showcase to stakeholders is impact**: A popular dissemination event at the end of a research project is a showcase, where "stakeholders" are invited to listen to the project academics talk about their findings. This is clearly not, of itself, impact. It could be that the stakeholders might take something away from the event and see how it might help them with their own goals, and it might become impact. But a showcase, of itself, is not.
- **Impact is something we can pay consultants to do:** This is something we will pick up later in the chapter, but there certainly seem to be views from management at various institutions that impact is something than can be bought in with consultants.
- **We can worry about impact at the end of the project:** This relates also to the "research showcase" view that once you have some research results, you then decide who needs to hear about it. In fact, in our experience, the best impact will be planned from the beginning of a project (if a project-based approach is being used for a case study) and in many cases there are elements of co-creation with stakeholders.
- **Academics should write their own case studies:** There is certainly a tension across the sector between those who think academics should be

responsible for all aspects of impact creation, writing, and evidencing, and those who think writing case studies is a management function. We clearly fall on the side of academics being in charge of their own impact, but that does not mean that every facet of the process of articulating a case study for research assessment should also be done by them. Research professionals can certainly help (and in our experience can be pivotal) in articulating the impact process and achievements, and frame the impact in assessment-friendly terms. This should be something done in partnership, working to each other's strengths.

- **Impact just happens, what do you need to plan for it?** The flip side of "we only do impact for the REF", there are some we have seen that seem to take the view that impact is something academics *should* be doing therefore there is no need for any support. We would, again, strongly argue that the best research comes where impact is planned as a fundamental part of the research process and a long-term goal of research activity, rather than assuming it will happen by osmosis.
- **We need to measure it every year so we know if it is getting better:** There is certainly, in some institutions, a management culture of "REF-readiness" that feels entirely output only, imposing large administrative workloads on academics so they can present their work to be judged by those who wish to measure it and see how it is progressing, often year on year with little support for those academics in the ensuing period. We would suggest that this metric-driven approach is one of the key reasons academics disengage with these processes.
- **My podcast/social media profile/media coverage/conference presentation is impact:** It is often the case that public communication is argued to be impact and that a popular podcast or a large twitter following are impact. They are not, although there are ways in which an academic can engage with an audience beyond academia and, as such, has potential to lead to impact. But without some form of significant change among stakeholders, it is simply communication.
- **I've been involved in a large, multimillion-pound, multipartner project, and there have been lots of publications and events**: There may be impact here, but unless the researcher is able to take ownership of their own role within this project, has understanding of the value they have brought to it, the outcomes of their value, the value that others have brought, and of how it relates to their own fuller body of research and impact, it is often difficult to take anything meaningful from this.
- **My paper is in a conference/<journal x> with an impact factor of <y>, therefore there is impact:** It is not, that is academic dissemination.

So where do these views come from? We will return to this later in exploring the views of our interviewees, but our own reflections are that this comes from a place where impact is imposed onto academics, rather than being seen as part of academic identity, and where knowledge is weak so, to use a popular analogy, in the kingdom of the blind, the one-eyed man is king. Or, to put it another way, if your own knowledge of impact is poor, why will you challenge the poorly informed views of a colleague or manager? So, instead, these myths perpetuate.

We have been critical of top-down/management approaches to research impact that does little to support the academic nature of curiosity and inquiry and instead has people from outside of the academic community telling academics what to do, with no means of being able to demonstrate they are relatable to academic practice. We will continue to be critical of this approach because it is our view that this is one of the fundamental reasons impact remains poorly understood and not embraced by academics. However, we should caveat this with a further observation (and this is probably why our more general observations are borne out of frustration) that we have worked with, and observed some inspiring and excellent research leaders throughout our careers (and indeed have interviewed two of them for this book). They lead from the front, being excellent research academics themselves, and can inspire and support because they understand academic identity. However, they do seem to be in the minority, and that was reflected by our conversations with interviewees.

There seems to be a view in some institutions that because academics are "allowed" by their employer to do research, they should be directed by the employer about how they should do it. We prefer to reflect upon the symbiotic relationship between employer and employee – an academic can be left alone or supported to do good research which, in turn, results in prestige and income for the institution (by way of quality-related (QR) funding coming from the REF).

However, we still see too much effort among university management and leadership to try to distil impact into a formula or a process model that need only be followed to elicit good results (even though there is little evidence to show this would actually be the case).

We can distil this process model into a number of key elements (Table 6.1).

In drawing this section to a close, we would hope that any reader would by this stage be convinced that impact, if successful, should not be top-down driven or managed with endless measurement and "peer reviews" (where peers lack the knowledge to appreciate the nature of the research/impact but that does not stop them having an opinion on it). Instead, we strongly argue that those most successful at impact do it as part of their academic identity, and have impact case studies ready for research assessment not because they have been "planning for REF" but because they see impact as important, have a research ecosystem that engages with stakeholders in an organic manner, and are good at it. We will explore this further.

Table 6.1 Impact Process Model Summary

Element	*Motivation*	*Reality*
Linear process leads to impact	Do some research, publish this research, identify stakeholders, disseminate to stakeholders, have impact	Doing good research, underpinned by good research culture, if impactful, is sometimes unpredictable, as is the nature of research
Stakeholder mapping	Who are the key stakeholders for this research and how can you reach them – perform a stakeholder mapping exercise to discover this	Good research culture, and good research support will result, if appropriate, in academics having good relationships with stakeholders and develop organically as the value of the research is articulate and further research is supported
Using the correct vocabulary	If you use the right words in a case study write-up, you will have good impact	Good impact comes from good research; while articulation is important, a magic formula or special words will not draw out good impact from poor research
Management oversight	Senior leaders need regular sight of impact cases to make sure they are going in the right direction	Unless senior leaders understand impact, and are providing support for impact, there is little point in this oversight and can feel like surveillance or interference
Regular scrutiny	Mock REFs/regular reviews allow an institution to see where the case studies of the highest potential are	Regularly measuring the same thing without incubating or supporting its development is detrimental to the motivation of the individual whose work it is, particularly when it is being judged by people who do not understand either the research or impact process

Impact as a Part of Academic Identity (For Some)

In Chapter 4, we presented a large volume of Freedom of Information (FOI) data that highlighted how ad hoc the approach to impact has been for a lot of academics in a lot of institutions. It is not surprising, therefore, that knowledge about impact remains so incomplete across the sector. In exploring our third major piece of empirical evidence, which we have touched upon in other chapters, we present further the key findings from interviews with several "impactful" academics, as defined in Chapter 1. In exploring these interviews, we are mindful that views expressed might be controversial or not reflect the views of their employers! As such, we have chosen to anonymise respondents and generalise responses, and not use direct quotes. As we have already stated earlier in this chapter, there was a great deal of consensus among our interviewees, regardless of institution and role. The interviews were semi-structured in nature and explored a number of key points:

- What is the nature of the impact you do in your research?
- Why do you do impactful research?
- What support do you have from your institution?
- What are the challenges you've faced in being an "impactful" academic?

We will present the more general issues that emerged from these interviews before looking in more detail as specific aspects of impact management and culture.

The Impactful Academic

One of the key issues that arose with the majority of our interviewees is that no-one planned to be an impactful academic. Their career trajectories evolved, sometimes unexpectedly, leading individuals into roles they hadn't been encouraged into. A number of those interviewed said they had evolved a more impactful mind-set, almost by chance, because either they were doing work they felt should be disseminated to stakeholders, or they had been approached by stakeholders to understand how their work might be applied. While this shows the unpredictable nature of career development in academia, where opportunities can arise from unforeseen circumstances or decisions, it also shows that individuals who can see merit in adapting to these new roles, because they can see value in it, were generally evolving professional interests and societal need in parallel. Furthermore, it shows the importance of those whose research ecosystem extended beyond academia into stakeholders, and how having a meaningful relationship with stakeholders resulted in both better research (something we will pick up in Chapter 8) and how individuals can work for mutual benefit. None of our interviewees described a linear process that followed the path of doing some research, then searching for stakeholders who might be interested in the results. For those we interviewed, impact and engagement with stakeholders was an intrinsic part of the research practice and, in a lot of cases, methods and data collection was co-created with stakeholders.

Complimentary to this was a lot of reflection on the evolution of their academic identity and growing in confidence that their approach is valid, reinforced in part by seeing the valuable impact their work could have on the stakeholder community. One interviewee, whose work centres on aspects of domestic abuse, felt empowered by being able to help victims of abuse to have more positive life outcomes, rather than just writing about the nature of abuse. The journey from being considered a junior in the field to gaining recognition and establishing oneself was a central theme and there were many discussions on the personal growth of an academic navigating through the early stages of its career, facing imposter syndrome, and striving to establish credibility within a competitive field. This evolution is not just about gaining experience but also about overcoming self-doubt and learning to value one's contributions to the field, and imposter syndrome was encountered a great deal. Not helped, in a lot of cases, by a lack of mentorship and, in some cases, explicit hijacking of this direction by line managers who could not see its value.

Personal stories within the interviews emphasise the intrinsic motivation behind pursuing impactful work, showcasing individuals driven by a desire to make a difference rather than by institutional rewards or aligning with faculty or institutional mission statements. They did this work because they believed in it, they could see its value, and in a couple of cases because interviewees said that they knew they were good at it and enjoyed it. What was clear from these discussions was that all interviewees believed that working with their stakeholders was an explicit part of their identity, and what it meant to them to be an academic. It reflected the personal satisfaction derived from impactful work if it aligns with one's ethical and professional values.

Regarding ethics, a number of subjects also brought up the ethical tensions between doing research with stakeholder communities because of the research potential it has, especially if they could see how the research could be shaped into a case study, and how it was executed to be ethical in nature and in the best interests of the stakeholders. For example, returning to the domestic abuse researcher, they were always mindful that their primary responsibility was to their research subjects not to their own research prestige. A number also noted that ethics committees at their institutions would rarely probe about impact and the ethical issues that might arise as a result, the focus was generally on what we might see at the traditional ethical screening of informed consent, data collection, briefing, etc. (Bayley and Phipps 2023). This issue probes the cultural norms within the academia and also places some practice at odds with the "impact industry" which focusses not on why we do impact research but how to do it and how to write about it, which a number of subjects observed can raise concerns around the overall integrity of impact work. For instance, one was told, post-REF2021 policies being published, that they should look to "work up" a case study, rather than being more broadly encouraged to do research that could help the stakeholder group.

In terms of personal research planning, it was interesteing to note that all of these researchers felt a close affinity with their research discipline. None of them were what we might refer to as "project researchers", who would move into different project areas where there were funding opportunities. Our interviewees were all very clear about their research discipline, which meant they could all build networks, data sources, and critical mass through long-term strategic planning. This and the execution of impact activities within the research discipline were crucial for achieving meaningful outcomes. This helped unpick the challenges of aligning personal academic goals with broader institutional or societal objectives and underscored the need for thoughtful, well-planned approaches to maximise the effectiveness and reach of impact work, rather than seeing impact as a bolt on at the end of a project you were only doing because there might be something "REFable" in it. All of our subjects were symbolic of a challenge to the traditional dichotomy between research and impact, questioning how academic value is assessed and rewarded. Subjects frequently debated the often-undervalued nature of impact work in comparison with traditional research outputs. In one case a very senior academic with very significant impact, and outputs, reflected on being told by their head of department that it was all very well doing "applied" research, but it needed theoretical depth too, as if the two could not work in a complimentary manner. A number of subjects, as was reflected in the anecdote in Chapter 1, were told that their work was viewed as second class by research managers until the REF announced that impact was something that was going to be measured and would bring funding. However, most saw little change in support at this time, just research managers telling them what to do, just more so, and more demands to justify the approaches they had been using (in many cases for many years with evident successful outcomes) to people who seemed to have little grasp of impact practice, something we have already explored in the previous chapter.

This was further reflected in many discussions around the complex interplay of dynamics within academic institutions, where bureaucracy, politics, and resource allocation can significantly impact one's career and the execution of impactful work. Subjects discussed the challenges of navigating these dynamics (or simply ignoring them because they did not believe their institution could provide anything helpful) and the often-frustrating reality of achieving recognition and support for impactful research. A significant challenge discussed was the scarcity of resources allocated to impact activities that could, in turn, stifle innovation and restrict the scope of academic work, at a time where they were being "advised" to "do more" impact by those in charge of research assessment in their institutions. It was interesting to observe that in the majority of cases, our interviewees did not know where the QR funding that their impact case studies in REF2021 produced had gone, and there was certainly, in the majority, no evidence of reinvestment of that money to the researchers who had generated it. This was once again used as a way to illustrate how institutions only care about outcomes, rather than supporting those who can achieve them (and perversely with investment could achieve even more). However, this was not always the case and in two instances we were told that at that given institution, all QR money was passed back to academics for investment in research.

In general, however, interview subjects were highly critical of senior leadership and management culture in their institutions. One commonly occurring theme was that senior management in their institution often operated reactively rather than proactively, which instilled little confidence that there was a critical mass of knowledge about impact policy at the top of the organisations.

This reactive approach, in the view of many, hindered the effective integration and support of impactful academic work, because, as stated earlier, there was a very clear view that good impact required strategic, long-term planning and the establishment of what some referred to as a research culture in their organisations (which, predominantly, was seen as lacking).

The management practices described by our interviewees often did not align well with the needs of the academics or how they felt that they were aligned to their research discipline. For example, the focus on short-term metrics and neglect of deeper, more substantive engagement with research impact led to many of our interviewees feeling unsupported and undervalued and simply viewed as useful because the management strategy for income maximisation required them to be cogs in the machine. Certainly, interviewees acknowledged the symbiotic relationship between institution and individual – they appreciated that they had workload allocation for research, and they could see that there had to be some return on investment. However, they appreciated less being told by those without an academic identity how to conduct research (in one case an interviewee was told by someone from the finance function of their university that they need to write "higher-quality papers" in order to maximise QR return – a surprise, we are sure, to be informed of this in such simplistic terms, as if they just needed to write better).

Many interviewees were openly critical of the quality of leadership within the sector, pointing out that high salaries do not necessarily attract leaders with the right capabilities and will naturally draw in those with a management, rather than academic or leadership, identity. This situation is exacerbated, we were told, by a culture where VCs and other high-level managers surround themselves with agreeable staff rather than those who would challenge poor decisions, leading to a lack of innovative or effective management and a leadership culture that could not lead, just instruct and demand. This, in turn, resulted in a critique of the impersonal and often bureaucratic processes that dominate the management landscape in academia, particularly around the REF preparations and impact assessments. These processes were viewed as overly mechanistic and not tailored to the actual strengths and needs of individual departments or disciplines – there was rarely a feeling of two-way support – more presenting ones research and impact to a panel of people who understand neither the discipline nor the research culture and, instead, would make judgements based upon fudged metrics, repeated impact tropes,[3] and subjective interpretation from a poor knowledge base.

While our interviewees didn't explicitly discuss "management surveillance" within universities, there was a great deal of discussion of management practices and oversight, such as how management monitors or controls on academic activities that could be viewed as aspects of surveillance and control. The overarching themes suggest a management style that may not fully understand or support the nuances of academic impact work, leading to practices that might feel restrictive or misaligned with academic goals. The concerns about management include a lack of proactive support for impact activities, superficial engagement with the requirements of impactful research, and a focus on bureaucratic processes that do not necessarily translate into genuine academic advancement or recognition. These points suggest a broader context where management might be seen as overseeing or regulating academic work in ways that do not necessarily align with the best interests of academic staff or the goals of impactful research.

Interviewees were critical of the impact processes in place within their institutions. It reflects a general dissatisfaction with how these processes were implemented, emphasising that the processes often seem disconnected from the actual academic base and are not effectively integrated into the broader research culture. For instance, they discussed how impact is often articulated as a linear process from output to dissemination to impact, which oversimplifies the complexity and depth needed in impactful research.

Furthermore, it suggested that there is a lack of deep understanding and support for impact within the management practices of universities. The processes were described as overly bureaucratic, with an emphasis on procedural compliance over genuine engagement with impact activities. The use of external consultants, and the greater trust in their voice than that of the academics in the institution, and a management-by-checklist approach are mentioned, contributed to a feeling among academics that impact activities are being monitored or managed in ways that didn't necessarily enhance the quality or effectiveness of the research impact.

One of the factors that often came up with interviewees regarding these processes was how "impact systems" were used by their institutions, who had invested considerable sums of money in software systems that would help them manage impact within the research assessment cycle. In general interviewees describe being "encouraged" to post outputs, evidence of impact, and testimonials from stakeholders that can be used to help qualify the scope and rigour of the impact, in central repositories for those with research assessment responsibilities to "review and assess" and, where needed, encourage the cultivation of further evidence.

Not a single interviewee saw the value in that. Given that impact was part of their academic practice and identity, they all maintained their own repositories of evidence and had good relationships with stakeholders, so generating testimonials was not a problem. Their view was that these systems were used as a power grab by their institutions, so that in the event of an academic moving to another institution, their previous employer could continue to claim their case study without their support (or alternatively have to pay to have the academic continue to support the case study submission). And many also viewed it as some form of monitoring and further attempts to measure what the institution did not understand. No-one saw any value in these systems. As one interviewee said "Why would I share my contacts with them, they'll just screw up the relationships I have with them".

Our interviewees, with a couple of exceptions, were clear that their senior leadership's engagement with impact was superficial, and an area in which they could rarely lead. This often resulted in impact being treated as an administrative checkbox rather than as a core academic endeavour. This superficial treatment could undermine the potential for genuine and transformative impact from academic research. Yet, as we pointed out frequently to our subjects, their impact had been reviewed by peers in their research disciplines (via the REF) as high quality. Ah yes, we were told, but that was due to the support of their peers and their stakeholders, there was rarely any encouragement from their institutions.

When considering these discussions with interviewees around management and leadership, we can see a clear correlation with the findings of the FOI results, which illustrate how the rhetoric about impact and research culture in institutions is not necessarily reflected in practice. Institutions do "things" but it's not convincing that they are doing the right things to best support their staff, who have already demonstrated their quality among their academic communities. The lack of institutional knowledge results in bad practice and, in some cases, negative impact on people's careers. Overall, the interviews paint a picture of a system that is more concerned with appearances and administrative processes than with fostering genuine and effective research impact.

These comments should be of concern to senior leaders in the higher education sector. These were not outliers or self-identifying rebels against higher education norms, they were all high-quality academics (as independently judged by academic peers) who felt their institutions were more of a hindrance than a help with their research identity. These critiques highlight a need for a more thoughtful and academically engaged approach to management within universities, particularly in how research impact is fostered and supported. Perhaps most concerning was how often the term "toxic academia" was used by our interviewees.

Toxic Academia

Toxic academia has been explored in both the academic literature and culture around higher education (e.g. Smyth, Smyth and Christie 2017; Pelletier, Kottke and Sirotnik 2019), and refers to a culture within academic institutions that prioritises productivity and institutional prestige (and the prestige of senior leaders) often at the expense of the well-being of its faculty, staff, and students. This environment is characterised by highly competitive practices, excessive workloads, and a relentless pressure to publish, secure funding, and achieve full-time contracts. Such conditions can lead to severe stress, anxiety, and a pervasive sense of inadequacy among academics (Smyth 2022; Erickson, Hanna and Walker 2023). The competitive atmosphere, reflecting the neoliberalisation of higher education in the 21st century, can also foster isolation rather than collaboration, where the success of the institution (and therefore the senior management) is highlighted over collective achievements and supportive academic relationships.

Moreover, toxic academia often perpetuates a cycle of exploitation, particularly of junior academics and graduate students who may face uncertain job prospects, low pay, and limited job security. This exploitative nature is compounded by a lack of sufficient mentorship and support, as senior faculty members may also be overburdened or incentivised to prioritise their own careers. The result is an environment where the academic ethos of inquiry and learning could be overshadowed by the pressures of survival within the field. Such toxicity not only undermines the health and morale of those within the academic community but can also stifle innovation and genuine intellectual growth, ultimately affecting the quality of education and research.

There were plenty of discussions among our interviewees around what they viewed as toxic academia within research and impact culture. They discussed several aspects of what might be termed "toxic academia", focusing on the challenges and pressures that academics face, especially around the themes of impact work, institutional pressures, and the ethics of engagement within the academic sector. These insights paint a picture of an academic environment where pressures to meet external metrics of success, lack of support, and ethical dilemmas contribute to a toxic culture. This environment can undermine the well-being of academics and the quality of education and research, highlighting the need for systemic changes in how academic work is valued and supported.

We have already touched upon concerns about the ethical dilemmas faced by those in the impact industry, highlighting concerns about maintaining integrity in a field that sometimes pushes superficial achievements over substantial contributions. However, there is clearly a toxic element to this, and some of our interviewees felt that had, on occasions, been pushed towards unethical exploitation of stakeholders in order to "improve" the case study. This reflects a broader issue in academia where there is a tension between producing quantifiable results for evaluations while maintaining rigorous academic standards.

There was a view that some of this toxicity comes from a culture of competition over collaboration, where the drive to secure individual or institutional prestige can overshadow the fundamental aims of education and scholarship. Again, under the shadow of neoliberal element of 21st century academic, the management belief in competition-driving individuals in their organisation was contradictory to the sharing of knowledge with peers, for a strong community and stronger impactful work, which are core to the academic ethos. Some felt they were "encouraged" to behave unethically in order for the institution to get the edge over institutional "competitors", a curious phenomenon of modern academia reflected also in the reluctance of many institutions in our freedom of information requests to share anything they felt might place their institution at a competitive disadvantage in a community which strived to judge the value of research and impact independently. And perhaps a reflection of the failure of management to appreciate the nature of research assessment.

Pressure and imbalance in workloads, a frequent facet of toxic academia, often arose in our discussions. Our subjects described scenarios where individuals are pressured to produce work that aligns more with institutional goals (or areas of "research strength" in institutional mission statements) than personal or academic integrity. This kind of pressure can contribute to a work environment that values metricated output over quality, leading to stress and ethical compromises.

For younger researchers, workload was aligned with career insecurity and instability. A number felt their early career years, post PhD, were met with demands for teaching (and as junior academics they felt they had little to bargain against these demands) yet at the same time being told to ensure they kept on publishing otherwise the fixed-term contract they were currently on would not transform into a permanent one. In one case, an interview subject said they were told if they had no time to write during the week, they should do so at the weekend if they were to "survive" at their institution. This instability is part of a larger narrative of academic careers being fraught with uncertainty, often exacerbated by short-term contracts and the pressure to continually secure funding.

Imposter syndrome was also brought up by interviewees too, and we can see this go hand in hand with toxic culture. One clear instance is seen in the segment where one of our interviewees doubted their knowledge and contributions relative to peers despite being engaged in significant work and contributing to academic conferences, they expressed uncertainty about their own intellectual contribution and sought validation through interactions with more established colleagues, many of who did little to reassure this impact path, saying it would be "better" to take a more traditional academic career path.

Furthermore, the interviews, on occasion, reflect the systemic and institutional aspects that can exacerbate feelings of imposter syndrome. There was mention of an academic feeling under-recognised and unsupported by their employer highlighting how environmental factors can intensify personal doubts. This was compounded by the individual's reflection on their role and contributions, questioning the value of their work in the absence of institutional acknowledgment. Such a context underscores how imposter syndrome is not only a personal internal conflict but can also be significantly influenced by the surrounding professional culture and lack of support, which can perpetuate an individual's feelings of being an imposter in their own field.

It was clear in some instances that some of our interviewees had experienced deliberate attempts to undermine their work and their academic identity, and comprise key facets of toxic academia. In a number of cases, this was senior academics in their institutions suggesting their approach to research was incorrect or the rather wonderful statement that seems to hang around research management these days: "I'm not sure that's REFable".

When asked why interviewees felt their managers and some senior colleagues did seem to be going out of their way to talk their work down, there were a couple of key thoughts – first, the "I didn't do this, therefore you shouldn't" mentality, where a "successful" senior colleague sees their career trajectory as the only way to be successful in academia, and therefore they develop a fundamental disinterest in any other route. And second that the vacuum of knowledge around research impact among much university management results in a fear of the unknown – it is better to dismiss something as irrelevant than try to understand why it might be useful. Which, conversely, returns to one of our key arguments – unless people understand *why* research impact is important, there will only ever be a shallow motivation to achieve it. Yet rather than try to understand it (and in terms of REF policy, all of the documents that set out what is required are publicly available for anyone who wants to read them) far better, we were told by interviewees to dismiss those who do it instead.

The Role of the Impact Industry

A final aspect of the discussion which is worth exploring in more detail is reflections on the impact industry – the ecosystem of consultants and trainers that has grown as a result of the increased monetisation of research impact and that we have already explored in the previous chapter. As an aside, it is interesting to note the growth of a number of consultants and training companies offering services around "building research culture" since UKRI/ Research England announce that measurement of institutional research culture will be a major part of the next REF.[4]

The role of consultants in delivering training, supporting, and assessing impact case studies was a frequent discussion point in our interviews. We draw reflections from our interviewees. It is also something we have had some interesting experiences with ourselves.

Once again one might view the (arguably excessive) use of consultants within research evaluation as an illustration of neoliberal tokenism within modern higher education. In an ideology that encourages the treatment of education as a commodity, consultants help academics and institutions increase the measurable impact of their research and "return on investment", rarely do they empower academics to develop their own impact as part of their academic identity.

The rise of consultancy roles in academia can be seen as a move towards outsourcing functions that were traditionally handled within academic institutions. This aligns with neoliberal policies favouring privatisation and a reduction in state funding for public services, including public higher education, and highlights a concept that an institution will be more likely to trust someone external they are commissioning for a service rather than turning to the knowledge of those within the institution.

The critique of the impact industry among our interviewees emphasised the ethical considerations and the superficial engagement by some institutions. Subjects discussed how some organisations use impact activities as a facade to enhance their image rather than for substantive change and how they were more likely to engage with impact consultant who could articulate and communicate their services in a manner that was more in line with what one interviewee called "LinkedIn language". There was a feeling by some that there was greater trust in the external consultants than academic staff in their own institutions with a proven track record of impact, and some felt that the engagement between managers and consultants led to a shallow engagement with impact culture by the institution – a view that management would bring someone in to do the impact bit, especially if their social media profile reflected they had a magic formula for research that would guarantee good research assessment. There was a view that there was a need for the balance between creating genuine societal impact and the pressure to perform impact activities for reputational benefit, and some questioned the integrity of practices within the industry.

There were varied attitudes towards the impact industry, with some academics suggesting that one-to-one work with a knowledgeable impact consultant had value and allowed the individual to build confidence that their work was of a high standard, regardless of the lack of motivation from their own institution. However, the discussion did highlight scepticism and the perceived necessity of the industry in academia by many in management, because they had no faith or trust in the academic staff they employ.

There were concerns about the real value added by the impact industry. It notes situations where institutions might bring in "big names" for talks or workshops that appear valuable but might not actually contribute significantly to the institution's goals. This suggests a view that the impact industry can sometimes engage in superficial activities that are more about appearances than substantive improvements. However, despite criticisms, there is also an acknowledgment of the impact industry's role in providing necessary services that institutions might lack internally. Such as offering expertise and resources that help academics and institutions achieve their impact goals. Some also viewed the impact industry as a facilitator that helps unlock potential in academic work, suggesting a positive view of its role when properly aligned with the goals of academic research and community needs.

Nevertheless, individuals in the text express resistance to engaging in what they perceive as the impact industry's more superficial or formulaic approaches, like generating clickbait content. This resistance was tied to a desire to maintain academic integrity and depth in their work and a concern that someone external who is not an academic peer will struggle to appreciate the deep knowledge in the work, so will instead revert to process and metrics to "assess" its quality.

One of the areas where the impact industry was seen as very visible was in impact training – developing the knowledge of impact within an institution by those external to it. There was general view that training and career development are not a core part of most academic experiences. A number detailed the path of an early career research/research training as:

1 Do a PhD (and receive mandatory training about ethics, publishing, and the process of achieving a PhD at the host institution).
2 Publish from the PhD.

3 Continue on your own, with minimal support from either institution or colleagues (many of who are in the same position). The main support received is to be asked "have you got any funding this year" during their yearly appraisal.

Certainly, many of our interviewees had experienced impact training and some of them saw value in at least getting a cross-sector of colleagues in an institution up to a certain knowledge level regarding impact. However, most view the impact training they had received in a context that highlights a lack of deep understanding and support for genuinely effective impact practices, it was more a case of "this is what impact is" and "this is how you do it". Interviewees suggested that impact training is often delivered as a "train the trainer" model, where professional services staff attend conferences or webinars and then relay the information (or the PowerPoint files that received as part of their attendance) they've gathered without a substantial grasp of the deeper strategic implications. This approach can lead to a superficial understanding of impact, which may not adequately prepare participants to implement effective and meaningful impact strategies in their work. One interviewee felt that the process of bringing in external people to deliver a one-size-fits-all approach is disrespectful to the independence of the research process and felt it exacerbated the management belief that it is indeed something that can be done regardless of research quality. One also felt that the money being spent on consultants (reflected in our FOI findings) might be better invested in supporting academics who could use that resource to develop deeper impact.

Consequently, the interviewees often implied that there is a significant discrepancy between the enthusiasm shown for impact activities, such as workshops put on by professional service teams, and the actual support for deep, critical engagement with what impact entails and how it can be integrated into research effectively. This gap indicates a need for more robust and critically engaged impact training that goes beyond mere presentation and encourages a deeper understanding of the complexities and nuances of achieving real-world impact through academic research. One observed a training model that perhaps best reflects this shallow approach to impact training:

- Why are we doing it? Because the REF asks for it and that generates income for your institution.
- How do you do it? Follow my process model.
- What do you do with it? Write it up using these keywords and a prescriptive writing style.
- And if you do all this, you will be an impactful academic.

We should stress that this is one person's experience with impact training, and we are certainly aware that impact training comes in many shapes and sizes and a lot is bespoke and can bring far more value (indeed, as discussed before, some interviewees felt more confident about their work as a result of speaking with external consultants). However, we also know that that sort of consultancy is expensive and many institutions are not willing to invest until (as discussed earlier) a reactive leadership decides too late they need to invest in impact.

One other observation from a few of our interviewees, as a result of our questioning about this, was that, even though they had a proven track record of impact excellence from at least one REF (and in some case since 2014), they were not asked by their institution to deliver training to other colleagues. We had no interviewee that said their institution turned to them for help in developing a better institutional knowledge about impact.

This is certainly something we have experienced ourselves. While Louise, in her professional role, does deliver internal impact training to colleagues, this has never been the case with Andy in his academic career. In one instance, reflecting the ridiculousness of institutional culture in not trusting the knowledge of their own staff ahead of external providers, the day after Andy attended training by external providers at his own institution about policy engagement, he travelled to the institution of the providers to deliver training on policy engagement.

Conclusions

In this detailed chapter, we have explored the nuanced journey of academics who, often inadvertently, become influential in their fields and develop an impactful element to their academic identity, emphasising the organic development of their impactful roles rather than through deliberate planning or institutional motivation. This emergence often occurs as academics engage deeply with their work and recognise its potential value to external stakeholders, leading to a symbiotic relationship that enhances both their research and impact. The narrative also highlights the transformative effect this has on the academics' identities, fostering a growing confidence in their approach and the legitimacy of their work, especially as they witness the positive real-world applications of their research.

Furthermore, we see a strong body of evidence that delves into the academics' reflections on the ethical dimensions of their work and the broader research ecosystem, critiquing the superficial engagement with impact driven by institutional mandates rather than genuine societal need and an understanding of why impact can be part of academic identity. This critique extends to the academic management practices that often prioritise metrics over substance, which can stifle innovative research and demoralise academics. Moreover, the discussions reveal a pervasive lack of institutional support that fails to recognise and cultivate the intrinsic motivations of impactful researchers, leading to a disconnect between academic endeavours and administrative objectives. Our evidence underscores the need for a more integrated and grounded approach to research impact, one that aligns institutional goals with the authentic interests and values of the academic community, but it also shows that, even without institutional support, some academics become impactful because they see it as core to their academic practice, rather than something that is imposed upon them by management and policymakers.

The chapter articulates an argument for re-evaluating the approach to research impact within academic institutions and emphasises the necessity of embedding impact organically into the fabric of academic research identity rather than treating it as a secondary or an imposed policy requirement. It advocates for a cultural shift in higher education institutions to foster a genuine appreciation and understanding of the role of impact in academia. This shift entails moving away from the prevalent management-driven approach that sees impact as merely a strategic asset for institutional rankings and assessments. Instead, impact should be recognised as a core component of academic work that intrinsically enhances the quality and relevance of research, aligning academic endeavours more closely with societal needs and real-world applications. Or, at least, those who choose this path to be supported in doing so.

We underscore the importance of supporting academics in understanding and pursuing impact as an integral part of their work and suggest that institutions need to provide more than just tokenistic support for impact activities; they should cultivate an environment where impact is valued and understood deeply, starting from the grassroots level of academic practice. This involves rethinking leadership strategies, where academic leaders are encouraged to support impactful research activities not just for compliance with external assessments like the REF but as a critical aspect of academic life that contributes significantly to knowledge, society, and policy.

In conclusion, the chapter calls for a redefinition of what constitutes success and prestige in academic work. It argues for a broader recognition of impact activities that goes beyond traditional metrics and assessments to include the transformative effects that research can have on society. By realigning academic incentives and support systems to emphasise the value of impact, institutions can foster a more dynamic and socially responsive academic culture. This would not only enhance the relevance and utility of academic research but also empower academics to engage more deeply with the challenges and opportunities of the contemporary world, ultimately enriching both the academic community and the broader society it serves.

We are not saying that impact should be done by all academics. However, we are saying that being impactful does not detract from academic identity or decompose those who do into elements the neoliberal machine.

Notes

1 However, in my experience in using these devices to engage with the policy development process, while they are useful it is unusual unless the evidence aligns with the already decided policy direction, which might go some way to explaining why the academic in this scenario hadn't bothered with these approaches. In contrast during the early stages of writing this text, with an alarming sense of déjà vu, I was

following an email trail of computer science academics discussing their frustrations regarding policy around AI regulation, particularly around safety, and the establishment of a new AI initiative that drew little from the computer science community. There was a great deal of discourse around the STEM (science, technology, engineering, and mathematics) knowledge in the civil service and a frustration that there was little research input into the initiative, including one contributor who stated that the academic community should be the ones influencing its direction "even though we are not being asked". Again, it was clear that this group of academics, many of whom carried out excellent research around the topic in question that would undoubtedly have been of value to the initiative, could neither understand why they had not been asked nor know how they could provide input.

2 www.computerweekly.com/blog/Computer-Weekly-Editors-Blog/Chancellor-Jeremy-Hunt-gives-the-magic-technology-tree-a-shake [Accessed June 2024].

3 Familiar tropes we hear a great deal include "You need more testimonials", "Is your underpinning research 2*?", "Have you tried engaging with more stakeholders?", "How many stakeholders have you reached?", and "You need more reach/significance". These tropes are rarely suffixed with "and let me help you with that".

4 www.ukri.org/news/ref-2028-people-culture-and-environment/ [Accessed June 2024].

References

Bayley, J., & Phipps, D. (2023). Extending the concept of research impact literacy: Levels of literacy, institutional role and ethical considerations. *Emerald Open Research*, *1*(3).

Erickson, M., Hanna, P., & Walker, C. (2023). Moving to teaching online: Moral injury, pandemic, and the toxic university. In *Beyond the pandemic pedagogy of managerialism: Exploring the limits of online teaching and learning* (pp. 13–31). Springer Nature Switzerland.

Lacohée, H., Phippen, A. D., & Furnell, S. M. (2006). Risk and restitution: Assessing how users establish online trust. *Computers & Security*, *25*(7), 486–493.

McGrath, K., Hendy, J., Klecun, E., & Young, T. (2008). The Vision and reality of 'connecting for health': Tensions, opportunities, and policy implications of the UK National Programme. *Communications of the Association for Information Systems*, *23*(33).

Pelletier, K. L., Kottke, J. L., & Sirotnik, B. W. (2019). The toxic triangle in academia: A case analysis of the emergence and manifestation of toxicity in a public university. *Leadership*, *15*(4), 405–432.

Smyth, J. (2017). *Toxic university*. Palgrave Macmillan.

Smyth, J. (2022). The making of bullshit leadership and toxic management in the neoliberal university. In Cruickshank, J. and Abbinnett, R. (Eds.), *The social production of knowledge in a neoliberal age: Debating the challenges facing higher education* (p. 357). Rowman & Littlefield Publishers.

7 Research Impact as Part of Academic Identity

Homo Academicus Impactus

Introduction

The majority of this book so far has been, perhaps unsurprisingly, empirically driven and we make use of considerable data to argue our points about the challenges of the why in impact. However, in this chapter we apply a more theoretical lens to our understanding, to help us shape what a better approach to impact literacy might be, while still being mindful of the considerable evidence we have collected about how academics who do impact engage with it, how research professionals both support it, and want to support it, and how institutions are substituting deep knowledge with processes and expenditure.

A consistent thread through these discussions is that there have been many arguments that impact is an imposition on academic and more specifically research ideology and practice (e.g. Watermeyer 2014; Chubb and Watermeyer 2017; Śliwa and Kellard 2021). However, those impactful academics that we have spoken to present an alternative view where they are engaging with impact not because they have been told to, but because it is important to them. It is, essentially, core to their academic identity.

When we reflect upon academic identity, what it means to be an academic, and what motivates academics to do what they do, we have seen from interviewees that the relationship in between institution, personal career goals, and discipline is sometimes a tenuous one. In many of the cases there was a view that the institution was their employer, but not their master. Far more faith was placed in peers and the academic community, rather than the managers and senior leaders in the institution. No wonder, therefore, that management demands to "do impact" were dismissed or responded to with disinterest. And for those impact pioneers who did view impact as part of their core academic identity, this was generally in spite of their institution than because of. A number of those we interviewed, reflecting our own experiences, viewed impact as a fundamental aspect of their research because the nature of inquiry resulted in research findings which were important to share with stakeholders, rather than because they had been told in a Professional Development Review that they should be doing some impact with their research. And, in a number

DOI: 10.4324/9781003454991-7

of these cases, the stakeholders with which they collaborated were not viewed as data objects, or potential funders, but as equal peers.

So, in our exploration of the challenges of seeing impact as part of culture, rather than a management or policy, imposition, we feel it is necessary to explore further what motivates academics and what being an academic means. As such, we consider, as many might expect, the work of Bourdieu whose seminal text, *Homo Academicus* (Bourdieu 1988), explores this in detail.

Homo Academicus

Bourdieu made significant contributions to the study of academic life, and his work focused on how cultural and social capital, along with academic field and *habitus*, played critical roles within the academic field. In *Homo Academicus*, Bourdieu applied theories of social practice to the academic world, and offered a detailed analysis of the French academic field in the 1960s. It was a detailed examination of the social structures and dynamics within higher education and research institutions in France, with broader implications for understanding academic life, its structures, and individual motivations which we come to with this exploration of research impact (Williams 2020).

Bourdieu argued that the academic field is not just a realm of disinterested pursuit of knowledge but is also a site of struggles for power, authority, and prestige, deeply influenced by external social forces; and his research and theories provide a framework for analysing the academic world that emphasises the relational nature of power and the importance of social networks and capital in determining academic success. It reveals the complex interplay between individual agency and structural constraints, shedding light on the often-invisible forces that shape academic life.

Of course, given its situation in French academia in the 1960s and 1970s, it is not without its criticism (e.g. Wacquant 1989; Fisher 1990; Skea 2021).

Critics argue that focusing solely on context-specific analysis may restrict the generalisability of his findings to other national academic systems or to the evolving landscape of higher education in today's globally interconnected digital age. They also suggest that Bourdieu's emphasis on the reproduction of social structures and elite habitus may overlook the agency of individuals within academia to transcend their social origins or to enact meaningful change within the academic field.

Moreover, changes such as the widespread accessibility of higher education, the emergence of digital scholarship, and shifts in funding models have significantly reshaped the academic landscape. These transformations potentially challenge or complicate Bourdieu's original analysis. Critics also contend that Bourdieu's work, particularly *Homo Academicus*, could have more explicitly addressed issues of gender, race, and diversity within academia. The role of these factors in shaping academic experiences and opportunities has gained recognition, indicating a need for expanded analysis beyond Bourdieu's initial framework.

Furthermore, while Bourdieu's critical examination of the academic field has shed light on profound issues of power and inequality, some critics argue that his perspective may overly emphasise the negative aspects of academic life. This focus might obscure the positive dimensions, such as the pursuit of knowledge, intellectual freedom, and academia's potential role in fostering social advancement.

Despite these criticisms, *Homo Academicus* remains a seminal work in the sociology of education and a pivotal reference for understanding the complex and often hidden dynamics of academic life. Certainly, we see lots of the issues explored by Bourdieu in our own experiences and also in our conversations with peers for this book, and we will draw upon these significantly in later chapters.

There are a number of key themes that are drawn out of Bourdieu's analysis that apply to our exploration and evidence:

- **The field of academic life:** Bourdieu conceptualises academic life as a "field" – a structured space of positions within which academics (agents) and institutions operate. This field has its own set of rules, practices, and hierarchies. Academics compete for recognition, resources (such as research funding), and academic prestige, all of which are crucial for advancing in their careers. It is Bourdeiu's view that the academic field is characterised by its relative autonomy (therefore, if impact is not recognised as something that will elicit recognition, resources, or prestige, why would they be interested in it?) but also influenced by external forces such as politics, economics, and social trends.
- **Cultural and social capital:** Bourdieu's notions of cultural and social capital are central to understanding how individuals navigate academic life. Cultural capital refers to non-financial social assets, such as education, intellect, style of speech, and dress. In academia, possessing cultural capital can influence one's ability to publish, secure funding, and gain prestigious positions. Social capital refers to resources based on group membership, relationships, networks of influence, and support. In the academic field, social capital is crucial for collaboration, mentorship, and the dissemination of one's work. Certainly, for impact pioneers their networks extend beyond academia to include a broad range of stakeholders who might also be useful for the furtherment of an academic's work, which might be used, in turn, to exert pressure on the institution for, for example, promotion opportunities. Indeed, some interviewees have said how the support of such stakeholders in, for example, an application for a professorial application does carry weight and influence should the stakeholder be viewed as significant. However, we would suggest that this view is not commonly held among academic colleagues.

- **Habitus:** Habitus is another cornerstone of Bourdieu's theory, referring to the deeply ingrained habits, skills, and dispositions that individuals acquire through their life experiences. In the context of academic life, habitus shapes how academics perceive and react to the academic field, influencing their research interests, writing styles, and how they interact with peers and institutions, and their sense of belonging within the field. One's habitus is crucial in appreciating the role of impact in one's academic identity, as we discuss later. A traditional habitus is unlikely to engage with an impact ideology, however that is not to say that all habitus is equal, and we have spoken to many academics whose habitus would undoubtedly recognise the role impact with stakeholders has in their academic life.
- **Autonomy of the academic field:** While the academic field is relatively autonomous, Bourdieu highlights how it is also subject to external pressures from the economic and political fields. This tension influences research agendas, funding priorities, and the overall direction of academic inquiry, and therefore the development of academic career and identity which potentially compromises the field's autonomy and the originality of scholarly work.
- **Prestige:** Prestige, according to Bourdieu, is not merely an acknowledgment of scholarly achievement or intellectual merit; rather, it functions as a powerful form of recognition that contributes significantly to an individual's position within the academic hierarchy. Prestige operates as a form of symbolic capital that is both sought after and wielded by academics.

While all of these themes can help us understand academic motivation and identity in the present day and, indeed, help us better understand the core questions central to the poor understanding of research impact by both institutions and academics, there are three aspects of this framework which are of most interest to us. There are the field, habitus, and prestige.

"The field", in Bourdieu's framework, more broadly is a structured space of social positions and relations, defined by the distribution of various forms of capital (cultural, economic, social, and symbolic) that the actors in the field possess. Each field has its own rules, logic, and stakes of struggle, where the positions of the actors within it are determined by their relative power or capital.

In *Homo Academicus*, Bourdieu applies this concept to the academic world, considering it as a distinct social field with its own unique set of rules, hierarchies, and power relations that help us understand the motivations of the actors (in this case academics, managers, and leaders) and allow us to explore how they operate. Some key aspects of Bourdieu's concept of the field as it applies to the academic field are:

- **Autonomy:** Each field operates with a degree of autonomy and has its own criteria for judgment and evaluation. The academic field is characterised by its pursuit of intellectual endeavours, knowledge production, and scholarly recognition. One might argue that the academic field is unique in how much autonomy academics believe and, in some cases, do, have. There is a disconnect between institution and academic individual which, while symbiotic, are not always mutually respectful. It is governed by specific types of capital, such as academic credentials, publications, and peer recognition.

- **Capital and power:** Within the academic field, actors (scholars, institutions, etc.) hold positions based on their accumulation of capital (and of course this results in importance of building capital as a primary motivating factor for academics). Cultural capital (knowledge, qualifications) and symbolic capital (prestige, recognition) are particularly significant. The distribution of these forms of capital creates a structure of power relations and hierarchies within the field.
- **Struggles and strategies:** Actors in the field engage in continuous struggles to maintain or improve their positions. In the academic field, this can involve producing influential research, gaining prestigious appointments, or securing positions in high-ranking institutions. These struggles are guided by the habitus of the actors, which influences their perceptions, actions, and strategies. If the habitus of academics does not contain research impact or stakeholder engagement as part of its core, it will not have strong influence on the practice of individual academics to build their capital and therefore career success.
- **Field dynamics:** The academic field is dynamic, with shifts in power relations, entry of new actors, and changes in the distribution of capital. These dynamics reflect broader social changes, as well as internal developments within academia, such as shifts in intellectual trends or funding priorities.
- **Interfield relations:** While fields are relatively autonomous, they also interact with and are influenced by other social fields. The academic field, for example, is affected by political, economic, and cultural fields through funding policies, political climates, and societal values about education and knowledge. This is perhaps as close as Bourdieu gets to exploring what we might view as contemporary research impact.

Given we are exploring motivations for engaging outside of the (academic) field and incorporating other stakeholders (or their fields) into one's personal field, to encourage the adopting of research findings to better benefit others, Bourdieu views academia not as an isolated realm but as a space continuously affected by external societal forces, which shape its practices, hierarchies, and distributions of capital. He discusses the concept of capital beyond just economic terms, including cultural, social, and symbolic capital and explores how these forms of capital are exchanged between fields. For example, political capital (influence, connections in political spheres) can be converted into academic capital (positions, funding) and vice versa. This exchange shows how external forces can shape academic priorities and structures.

While each field possesses a degree of autonomy with its unique norms and hierarchies, they are not isolated and often engage in complex interdependencies. Bourdieu highlights that these relations involve the conversion of various forms of capital (cultural, social, economic, symbolic), and are characterised by power struggles that impact academic practices and priorities. Academics may use their scholarly capital to influence outside fields, just as external forces may shape academic agendas and standards. These dynamics are further mediated by individuals' habitus, which influences their ability to navigate these interfield relationships, leading to continual reshaping of academic structures in reflection of broader societal conflicts and changes, which might be influenced by external factors such as policy positions and neoliberal restructuring of academia.

While *Homo Academicus* doesn't directly address neoliberalism, the themes explored in the text have significant implications for understanding the effects of neoliberal policies on higher education globally. To relate Bourdieu's analysis to neoliberalism, we should consider the broader shifts in governance, economy, and society that neoliberalism promotes, and we have explored from the higher education perspective in the early chapters of this book, and how these shifts impact the academic field. In response to these changes, Bourdieu's work encourages a reflexive stance among academics and highlights the importance of collective action to challenge the neoliberal logic and advocate for an academic field that prioritises intellectual rigor, critical inquiry, and social responsibility. There is tension between the neoliberalisation of higher education and Bourdieu's work, and its analytical framework is highly relevant for understanding and critiquing the impact of neoliberal policies on the structures, practices, and values of contemporary academia, and also, as we have already discussed, highlighting why these top-down and management-led approaches rarely gain much traction among the academy as a whole.

However, as we have discussed earlier when considering the influence of, for example, civic university concepts, he also argues that heteronomy can manifest when economic or political imperatives dictate research agendas, the allocation of resources, or the structure of academic institutions. We might argue that heteronomy is a growing pressure in neoliberal higher education. For instance, as we have discussed earlier, when government policies prioritise certain research areas for funding, this can shift academic focus and resources towards those areas, potentially at the expense of less-favoured disciplines.

Bourdieu also focuses on the struggles within the academic field between those who align with external forces (e.g. those who adapt to market or political demands) and those who strive to maintain academic integrity and autonomy. This is a particularly interesting distinction given our explorations in this book, and it is certainly something we have observed ourselves and with our interview subjects – the tension between, for example, the "applied" research compared with theoretical colleagues. It is rare, in our empirical interactions for this book, that the theoretical researcher will view applied research as sufficiently academic or, in some cases (as articulated in interview) it is dismissed as a betrayal of academic purity.

Moving on from the field, Bourdieu's theory of habitus offers another useful tool through which to analyse various social and institutional contexts, including academia. Habitus refers to the deeply ingrained habits, skills, and dispositions that individuals acquire through their life experiences. These are shaped by their historical and social contexts and, in turn, influence their perceptions, actions, and opportunities. When applied to academia, Bourdieu's theory can help explain several aspects:

- **Academic culture and practices:** Habitus influences how academics interact with their environment, shaping their academic practices, research interests, and methodological approaches. Scholars often develop a scholarly habitus that aligns with the norms and values of their disciplinary communities (i.e. they are more loyal to their peers than their institution or their line management). This includes preferences for certain research topics, methods, and ways of communicating knowledge. External policy influences places this practice in tension, because the individual feels they should remain faithful to their discipline first.
- **Field dynamics:** Academia, as a field, has its own set of rules, power structures, and capital forms. Academic habitus shapes how individuals perceive and engage in these struggles, influencing everything from research priorities to the informal politics of departmental life.
- **Inequalities and exclusion:** Bourdieu states that habitus can lead to unconscious biases and exclusionary practices within academia. Those whose habitus aligns less well with the traditional academic culture – perhaps because of differing socio-economic backgrounds, gender, ethnicity, or educational pathways, or where they view applied practice or dissemination outside of the academic – might find it more challenging to succeed. Such misalignments can lead to experiences of marginalisation or imposter syndrome, something we certainly explored with several interview subjects with frustrations at the lack of acknowledgement of the importance of impact by peers, or a pride in being different to the "typical" academic.
- **Change and innovation:** Habitus also affects how change is perceived and enacted within academia. Those with a more conservative habitus may resist new ideas or innovative practices (e.g. the importance of impact as part of research assessment) that deviate from established norms (research practice underpinned by income and academic outputs), while those with a habitus more open to change may drive transformation in academic disciplines or institutions (are our impact pioneers evidence of a more open habitus to challenge the academic status quo?).

Overall, Bourdieu's concept of habitus helps us to unpick the motivations of academics and explain why well-established practice is sometimes resistant to change, particularly when that change is viewed as being imposed by external factors (whether that is management or from other fields). Habitus helps explain the continuity and change within academic life, the reproduction of inequalities, and the varying levels of ease and success individuals experience in academia. It also helps us unpick why impact pioneers have less discomfort with impact agendas (and in some cases dismiss the policy perspective on impact as management catching up with their approach to academic work) as their habitus aligns with stakeholder interaction and recognises the transfer of research findings to facilitate impact as a mutually rewarding practice.

In parallel with the established cultures and the habitus of academia, something that strongly manifested in our interviews, which is borne out with Bourdieu's theoretical framework, is the importance of academic prestige. Bourdieu's analysis of prestige in *Homo Academicus* challenges the notion of academia as a purely meritocratic sphere and helps explain why academics are, arguably, sometimes difficult to manage! Instead, he portrays academia as a complex social field where symbolic capital, in the form of prestige, plays a critical role in shaping academic careers, motivations, and hierarchies. Through this lens, Bourdieu offers a nuanced understanding of the power dynamics at play within academia, shedding light on how social and cultural factors influence academic success and the distribution of authority.

Prestige is accumulated through various means, including but not limited to, scholarly publications, awards, and positions held within prestigious institutions (and this has been explored by other scholars such as Blackmore and Kandiko 2011, Kwiek 2021, and Fyfe et al. 2017). It is distributed unevenly across the academic field, with certain positions, disciplines, and institutions holding more prestige than others. Therefore, academic motivations are not always aligned with salary or institutional goals, but with their own prestige motivations, which can, conversely, align with established habitus. Prestige acts as a form of symbolic capital that can be converted into other forms of capital, such as social capital (networks and connections) and perhaps economic capital (through higher salaries or access to funding). The distribution of prestige is closely tied to the *mechanisms of reproduction* within the academic field, which perpetuates the status quo across generations and can be resistant to change. Prestigious positions often go to individuals who already possess a certain level of symbolic capital, reinforcing existing hierarchies and inequalities. This self-reinforcing cycle perpetuates the stratification of the academic field.

Conversely, academics can experience struggles for recognition and prestige. These struggles are not just about personal ambition, they can be deeply embedded in the structure of the academic field itself. Success in these struggles often depends on mastering the unwritten rules of the field and effectively mobilising one's accumulated capital or perhaps aligning with the more traditional habitus, rather than being viewed as a progressive (or, perish the thought, an applied academic) and therefore a poor fit. Furthermore, the criteria for achieving prestige can vary significantly across different academic disciplines. What constitutes valuable contribution in one field might not hold the same weight in another, reflecting field-specific logic and criteria for evaluation.

Homo Academicus Impactus

Drawing from both our own observations and those of our interviewees, we can clearly see how the management culture that besets research assessment, and the reliance on solutions and processes does little to support those who actually conduct effective impact. While these approaches, and the use of external consultants, might help boost confidence of management that they are ticking the impact boxes, it perhaps shows more about the lack of knowledge (and confidence) by senior leaders and the need for external validation.

In contrast, what we draw from considerably with the interviews is that our subjects did impact because they enjoyed it, because they think it is valuable, and because they felt it part of their academic identity.

In the first half of this chapter, we have explored the nature of academia, and academic identity, and explored theoretical positions that argued impact is something that has been imposed upon academics, and therefore should be rejected. We made considerable use of Bourdieu's framework of academia to explore what it means to be an academic which, though written many years ago, still has many recognisable features to both us and those we interviewed, and we argue that unless impact is something that is adopted by the academy, rather than imposed upon it, there will always be considerable resistance to its value and it will rarely be anything but something academics feel obliged to do because research assessments (and to a lesser extent funders) demand it, rather than because they see it as something they feel is part of their research ideology.

However, what our own experiences initially showed and interviews helped us understand is that this was not just something we felt, that there are academics engaging in impactful research not because of research assessment and management pressures, or superficial career goals, but because it is a core part of their academic identity.

Academic identity among our interviewees was, at least partially, defined through engagement with impact activities. The individuals discussed seeing their involvement in impactful research not as a professional requirement but as a means to enhance the practical relevance and societal benefit of their work. However, there were also tensions evident between traditional academic roles focused on teaching and theoretical research, and the newer demands for impact and engagement with external stakeholders. This conflict influences how academics perceive their roles – while most adapted to incorporate impact as a core element of their identity, because they understood why it was important, some felt resistance by their colleagues and leadership as a dilution of traditional scholarly values.

We also saw how institutional strategies and policies around impact influence individual academic identities. Academics are not only shaped by their disciplinary norms but also by the strategic directions of their institutions, which can either support or constrain their engagement with impact activities. Those we interviewed who had encouragement from peers (and indeed stakeholders) were generally more confident to push back at institutional demands and metrics than those who did not feel external validation was valid given the discouragement they had received from those within their institutions. We would suggest that recognition and validation through successful impact activities can significantly enhance an academic's sense of self-worth and identity within the academic community. Conversely, a lack of recognition and support for impact activities can lead to feelings of marginalisation or questioning of one's role in academia.

Perhaps the strongest indicator of this was interviews with individuals we might refer to as impact pioneers, those academics who were doing impact before it was called *impact*! In these cases (and this was certainly reflected by Andy's scenario in Chapter 1) there was no encouragement from their institution to do this work whatsoever, unless the "impact" meant they could get money from stakeholders to fund research projects. However, these individuals still worked with stakeholders and felt personal reward in doing so. Their academic credibility against traditional measures of outputs, income, and peer esteem was not watered down as a result of these activities, indeed these activities helped build more robust data sets which resulted in strong outputs and great stakeholder interest in their work. For our impact pioneers, doing impact was fundamental to their identity as academics. In conversations with these pioneers (three within our interview cohort) they all said very similar things:

- I did it because I thought it was important.
- I did it because I could see the impact of the work on stakeholders.
- I did it because I enjoyed it.
- I did it because I felt appreciated.
- I did it because I was good at it.

All of these factors were personal motivations. There was no one saying if they adopted an impactful approach they would get promoted, and they rarely received praise from managers for doing this ("wasting time that could be spent writing outputs or bids"). But they did it anyway. And when the REF2014 guidance for assessing impact was published, they all attracted considerably more attention from their senior leaders.

Returning to Bourdieu's framework of academia, we would argue that the impactful academic identity is fundamental to what it means to our participants to be academic. While prestige, and the nature of prestige capital to these academics, is fundamental to understanding the intrinsic embedding of impact in their identity, there is also much to be drawn from Bourdieu's concept of the field and also habitus for these academics.

Bourdieu's concept of the "field" describes the distinct social and cultural arenas in which people and institutions interact and compete for resources and power. Each field operates according to its own set of rules, norms, and stakes, and is relatively autonomous from other fields. Examples he explores include the fields of art, education, law, and politics. In each field, actors (individuals or institutions) possess varying degrees of capital (economic, cultural, social, and symbolic) that they use to maintain or enhance their position within the field. The structure of a field determines the forms of capital that are valued, and the struggle over these forms of capital reproduces the existing power structures within the field. Bourdieu's analysis highlights how cultural practices and preferences are tied to power relations and are instrumental in the reproduction of social hierarchies.

In Bourdieu's application of the field to academia, he views it as a competitive domain where scholars compete for resources, recognition, and influence. Within this academic field, the agents (scholars, institutions, etc.) possess varying degrees of this capital – cultural, intellectual, social, and symbolic – that dictate their power and status. Bourdieu describes the academic field as governed by specific rules and norms that define legitimate knowledge and scholarly practices. Success and prestige within this field are determined by one's ability to amass and deploy capital effectively, often evidenced by publication in top journals, securing of grants, and accumulation of peer recognition. This competitive landscape not only influences individual careers but also shapes the overarching priorities and intellectual directions of research. Given that the history of academia has not explicitly explored research impact as a form of prestige, we can see with this concept how policymakers might struggle to impose the importance of impact upon the actors in the field.

However, we would argue for those who see impact as part of their academic identity, the field extends beyond the ivory tower (a problematic and reductionist term that we do not particularly agree with, but a useful metaphor in this context). While we are uncomfortable with this term, it has some merit in this exploration about the boundaries of the field.

Certainly, our interviews highlight the competitive nature of academia, particularly how impact is pursued, valued, and recognised within different universities, and by different actors within universities. However, the impactful academic sees impact itself acts as a form of capital that academics strive to accumulate to gain prestige and improve their position within the academic field. This is evident in the discussion about the Research Excellence Framework (REF) and how institutions and individuals navigate its requirements to boost their status or secure funding. Arguably, impact has more prestige because it is now valued by academia, but there are some within the field who saw a more intrinsic value by extending the field to stakeholders with whom the academics interacted.

Bourdieu's concept of *illusio* (Taksa and Kalfa 2015) refers to the deep investment and belief that individuals have in the stakes and practices of a specific social field, driving them to engage in "the game" and recognise its value. For impactful academics, they have engaged with deep investment in the impact concept because they believe it to be important and see its value (and it is recognised by stakeholders).

There is also a tension between autonomy (the degree of self-determination within the field) and heteronomy (external influences) that is a crucial part of Bourdieu's field theory. Our interviews, and our own experience, show how external factors, such as funding bodies and government policies, but also the value placed in the research artefacts by stakeholders who can make use of the knowledge generated, heavily influence academic practices and priorities, and can, as is the case with our subjects, lead to shifts in research practice and academic identity.

Pierre Bourdieu's concept of "habitus" refers to the deeply ingrained habits, skills, and dispositions that individuals acquire through their life experiences, particularly through early socialisation in the family and education. The habitus is both a product of social conditions and an influence on individual practices and perceptions, operating beneath the level of conscious thought and actively shaping people's behaviour and responses to their social environment. It is durable, yet adaptable, and manifests as a "sense" or "feel for the game" in various social contexts, guiding actions without strict rules. This system of dispositions enables individuals to navigate their social world but also tends to reproduce existing social structures and class conditions because it is shaped by the individual's past and present circumstances, which are themselves influenced by broader social structures. Thus, habitus is central to Bourdieu's explanation of how social order is maintained and how social practices appear natural and self-evident to those engaged in them.

Habitus offers a critical lens through which to view the behaviours and practices prevalent in academia. This set of ingrained dispositions and cognitive frameworks, shaped by one's social and educational background, directs the ways in which academics perceive, react to, and navigate their professional environments. In academia, habitus influences research interests, methodological choices, and pedagogical approaches, often aligning with the prevailing norms and values of the academic field. This alignment helps individuals to "fit in" and succeed within their respective disciplines, but can also lead to a replication of established academic cultures and power structures. Through habitus, Bourdieu illustrates how academic practices are not just individual choices but are deeply influenced by social and historical forces, perpetuating continuity and sometimes inequality within the academic field. Therefore, as with the field, if academics cannot recognise impact as part of the academic habitus, it is unlikely to be seen as something important to their identity.

For impact academics, we would argue, the habitus is shaped with the influence of those outside of the academic traditions, which are embedded in a habitus which aligns more with the academic peers than the traditional management and leadership structures in their own universities.

The habitus of the individuals in our interviews is evidently shaped by the academic environment, which includes expectations around research, impact, and REF performance. For instance, the focus on achieving high REF scores influences behaviour, driving academics and institutions to align their activities and priorities to meet these externally set benchmarks, as institutions can exert pressure on academics (i.e. "unless you do what we instruct you will not gain promotion") to make them align with policy. However, it is also shaped by external stakeholders, who allow the academics to recognise a different form of prestige beyond the research assessment frameworks to have something of more genuine value.

The career choices and trajectories of the academics who took part in the interviews also reflect the influence of habitus. Their decisions on what research directions to undertake, how to engage with impact activities, and even how they perceive their own roles within the academic community are outcomes of their learned dispositions and the cultural norms of their specific academic fields but also by extending their discipline to incorporate stakeholders. They feel there is a deeper research culture as a result of strong relationships with stakeholders who, in turn, feedback both value and also research opportunities.

The interviews also highlight a variation in how different academics understand and engage with the concept of impact, which can be seen as a manifestation of their habitus. This includes their approach to stakeholder engagement, the value they place on different types of research outputs, and their attitudes towards the managerial aspects of academia. Most of our interviewees saw personal motivation, and stakeholder value, as more important than "pleasing" their institutional leaders, once again highlighting a breaking out from a more traditional habitus while at the same time still have a deep understanding of what it means to be an academic for them.

Arguably most significantly, if we are using Bourdieu's framework as a means to understand how *some* academics have incorporated impact as part of their core identity as an academic, whereas others have resisted something they see as an imposition on academic tradition, we come to prestige.

Bourdieu argued that prestige in academia highlights how academic capital is accumulated and recognised within the academic field. Prestige in academia is traditionally conferred through recognition by peers, citations, publication in prestigious journals, and appointments at esteemed institutions. Bourdieu argues that the struggle for prestige, which often correlates with access to resources and influence, shapes academic practices and priorities, influencing what is researched and published. This competition for prestige not only affects individual careers but also perpetuates systemic biases and gatekeeping within the field, often marginalising unconventional ideas or scholars from less prestigious backgrounds. However, for some academics, including those we interviewed, prestige can be extended beyond these traditional measures and incorporate prestige generated from external stakeholders which, conversely, conveys great academic capital now that institutions are being driven to incorporate impact into their funding.

Our interviews align closely with Bourdieu's notion of symbolic capital, which encompasses the prestige, recognition, and influences individuals or institutions accumulate within their field. Certainly our interview subjects all had the trappings of traditional academic prestige, such as publications, citations, and income. However, they also had something else. In engaging with stakeholders through networking collaborations and being viewed as either a thought leader or influential participant in the wider academic field, their prestige is recognised by their institution, who understand that these activities can result in between performance in research assessment.

Nevertheless, we can also see that most of our interviewees view the actual act of engaging with impactful activities and seeing the results of engagement with stakeholders as a highly prestigious outcome for their personal esteem, arguably in their view more important than the increased interest their institutions now take in them since the REF requires evidence of research impact from units of assessment within an institution. As such, we can see that while the external value of prestige generated from impact is important, there is a deeper, more personal prestige, and prestige from a wider ecosystem of peers that they consider more important.

Perhaps the most significant illustration from this work of how the alignment of impact with academic identity results in a deeper personal connection with impact came as a result of one of the discussions with an interview subject. In this case this individual was a senior professor in a Russell Group university who had previously worked at a different institution. Their research discipline is contemporary in nature, and we were aware of their work and how impactful it was, informing national and international policy, and working with key industrial players in the area. However, we wanted to explore with them why they had never submitted an impact case study, given the significance and reach of their work.

In unpicking this, we saw clear evidence of both toxic academic and poor knowledge exchange at their previous institution, where they had been told first by a senior manager, whose knowledge of impact did not strike as particularly informed, that their unit of assessment would not be interested in "your sort" of impact. Furthermore, a research professional in the institution had articulated a very basic, linear model of impact where they wanted them to identify a specific article that they could build an impact case study around. The academic viewed this as particularly problematic for them given their volume of output and research critical mass, all of which informed the impactful activities they did.

However, since unpicking this at interview, the academic in question clearly sees that what they are doing is impactful, and is working, for the first time, in building an impact case study underpinned by significant evidence and testimonials from major policy and industry actors.

Or, to put it another way, when they recognised research impact as part of their academic identity, they saw its value.

Impact as Part of Academic Identity – Andy's Case Study

By way of example, we can apply this thinking to Andy's case study and broader experiences, used as the introduction to this exploration of academic identity in this book. We can similarly align this with the broader concepts of academic identity and motivation driven not through external policy and management demands but because he believed it to be the right thing to do with his research.

Throughout this book, where we set out to try to understand why impact has not become a strongly embedded part of research policy and practice among our universities, one of our fundamental arguments is that while impact is viewed, as it is by some, as something that is imposed on academics by policymakers and senior managers, it will never be respected by academics as something that can (but does not have to) be part of their academic identity. In exploring this view, we have leaned heavily on some of Bourdieu's work on academic identity and the nature of academia, and a theoretical framework that both explains this position and helps up develop arguments that if impact can be part of academic identity (and this is reflected by the impactful academics we have interviewed in this project), it becomes something academia can own, rather than something that is forced upon them for the benefit of institutions but not individuals. Central to this exploration have been concepts of the field, habitus, and prestige, alongside that capital that exists in these spaces to help an academic realise their internal worth and also the respect of their peers. We can apply these concepts to this work to show how an approach such as the one described in this case study is deeply personal and embedded in personal academic practice.

Bourdieu's concept of the field (Thomson 2014) refers to structured social spaces or systems where individuals and groups vie for different forms of capital (social, cultural, economic, symbolic) according to specific rules and power dynamics inherent to each field. Fields operate semi-independently, each with its own norms and values, influencing and being influenced by the participants' actions and the distribution of power within them. Applied to academia he examined it as a competitive space where scholars vie for intellectual dominance and resources, such as research funding and academic prestige, governed by specific cultural norms and values. In this field, the academic capital (like credentials and publications) influenced one's ability to shape disciplinary boundaries and attain positions of authority within a scholarly community.

The case study and wider practice illustrates the concept of the "field" by exploring the dynamic interplay between academic research and societal impact, particularly, in this case, through public policy and practice on online safety. This case study, it could be argued, extends across several fields to incorporate others and illustrates the ongoing negotiation and application of various forms of capital (cultural, social, and symbolic) across these fields. It is important in these fields to maintain the stakes and rules of which they comprise, and in academic terms knowledge and impact are valued differently (and in some cases in academia impact is viewed not as capital but as external to the field). However, by embedding evidence-based practices using methodological rigour into educational policies, this case extends a particular academic field in a manner that is both academic but also has broader prestige.

The accumulation of capital in this case is evident. There is a cultural capital of expertise and knowledge that clearly has value to those who wish to improve outcomes for young people in addressing online harms, there is social capital from his networks and connections within the policy and educational sectors, and the symbolic capital from recognition and legitimacy-gained impact work that has been demonstrated to be successful, both in the goals of policy change and also the wider goal of peer-reviewed impact. These forms of capital influence both the academic and policy fields, altering perceptions and priorities within these spheres and shaping policy. Furthermore, the conflicts and challenges that arise from attempting to present a progressive position in how to address online harms is an illustration of the typical struggles and competitions for capital that Bourdieu emphasised, which in turn highlights how different forms of capital are essential for navigating and succeeding within any given field.

The work also illustrates the tensions between the pursuit of theoretical knowledge and the drive for practical applications, of itself a common struggle in academia. This tension is a site of struggle where different forms of capital (theoretical vs. practical applicability) are contested and valued. Through efforts to align research with tangible societal outcomes, this case study at least attempts to navigate these tensions but also works to redefine what is valued in the academic field, at least in his own field. A consequence, but not a motivation, for this redefinition, is the success as reflected in the case study in the previous REF with its partial focus on real-world impact.

Furthermore, the interaction between the fields of academia and public policy is significant in this case study. It allows the application of rigorous academic research to address practical issues in policy and education, influencing teaching practices and policy implementations in schools. This cross-field influence is pivotal in shifting educational norms and integrating new safety protocols, which are direct applications of his research findings. Such efforts demonstrate the fluidity and interconnectedness of fields as the work impacts broader educational and safety practices.

In summary, this is a potential model of how an individual can operate effectively across multiple fields by leveraging and accumulating various forms of capital that is valuable to them. The ability to influence educational practices and policy changes through his research exemplifies Bourdieu's notion of the field as a complex space of power dynamics, where capital is both a tool and a prize of the academic and policy-making arenas. Through strategic engagements and the application of research findings, this case study illustrates the notion of fields and also shows how one can shape the fields in which one operates if value and prestige can be generated in the application of research findings.

While the field might represent the external reach of the work, we can equally see how the research shapes on individual habitus that extends beyond traditionally held prestige. Habitus, as Bourdieu described, is a system of durable, transposable dispositions that individuals acquire through their life and career experiences. These dispositions influence perceptions, appreciations, and actions, shape an individual's worldview and their interactions with social structures. Again, we can posit a view that impact fails to gain traction in academic practice if it does not form part of academic habitus.

In the context of the chapter, this case study, and its underpinning long-term engagement in the field of online safety and the interactive approach to engagement with stakeholders reflects the development of a specific academic and practical habitus. This habitus is not just a personal framework that could be seen as not fitting a particular academic or external identity, but is one that is shared and reinforced through interactions with other stakeholders in the field, such as policymakers, educators, and the community at large. This habitus, informed by over 20 years of research and engagement, directs actions and research focus towards impactful outcomes, emphasising evidence-based policies and the importance of addressing real-world problems.

Bourdieu's work also discusses how habitus interacts with the field, the social arena in which struggles or manoeuvres take place for positions of power or resources. In this case, the academic field and the broader field of public policy on online safety are the arenas where the research can be applied. The stakeholder engagement described in the chapter exemplifies how this habitus and the field interact: research influences policy changes and educational practices, and feedback from these applications further informs research directions and methods.

Moreover, the concept of capital – social, cultural, and symbolic – plays a role in how this work is perceived and its efficacy in broader contexts. The work is not lacking in academic rigour, it stands up to traditional forms of academic capital (such as the peer-reviewed outputs that underpin the research) but the practical application of his research, and the ability to communicate and collaborate effectively with various stakeholders, enhance his social and cultural capital within the field of online safety.

Thus, this work can be seen as a manifestation of his habitus, interacting dynamically with the field of online safety to produce research that is not only academically sound but also practically impactful. This interaction between internalised dispositions and the external social structures exemplifies the practical application of Bourdieu's theoretical framework to understanding individual and collective actions in specific social contexts.

And finally, while we have already touched upon prestige in this discussion, there are a number of explicit examples of how prestige is shaped with this personal habitus. First, the long-standing engagement and the breadth of research across the field confer academic prestige. This is evident from a position where the research findings are used as reference points in discussions on online safety, where findings are utilised to shape educational and policy frameworks. This policy shaping is a reflection on the esteem the findings have among peers and stakeholders. The collaboration with policymakers, educators, and legal experts suggests a high level of social and symbolic capital. The ability to engage effectively with these diverse groups and to be invited to contribute to policy discussions and educational reforms is a testament to this prestige.

Conclusion

Exploring the nature of academic identity at a theoretical level and applying this to those we have interviewed about their own impactful identities, and our own work, helps us more clearly understand the tension between the belief by policymakers that impact is important and the resistance by academics to do it. Put simply, they do not see it as part of their identity, their field, or their habitus and, given it is perceived by many to be imposed upon them simply because "the funders" or "the government" demands it, does not result in engagement but resistance.

However, those who have carried out impactful research because they think it important, not because of the imposition, and in a lot of cases in spite of management steer inside institutions, the connection to impact at a policy level is far more clear. Nevertheless, as we have explored in this and the previous two chapters, there is still a breakdown between policymaker, management in institutions, and academics themselves, due to a failure of senior management and policy (in the view of our interviewees) to both understand what impact is at an ideological level, and whose motivations are not academic but financial.

To use Bourdieu's vocabulary, if we, as academics, do not see the prestige of impact or how it aligns with our habitus, a top-down policy that attempts to impose upon and reshape this will invariably be resisted. Both research professionals and academics we have interviewed have observed frustrations with the way impact did not matter until there was a financial reward for doing so, yet when those "in charge" did engage with it, their influence was rarely one of leadership because they did not have knowledge of either the policy expectations around impact or a deep appreciation of impact. In a lot of cases these academics did indeed see engagement with *their* stakeholders and the subsequent effect, change, or benefit that results, and they did impact for academic satisfaction, not because they had been told to, and they did not appreciate "advice" from those who had, until recently, questioned why they were doing it at all.

Interviews with impact pioneers showed that they did impact because they believed it was important to them, because they enjoyed it, and because they were good at it. Stakeholders appreciated what they had done, and saw its value, which, in turn, raised the self-worth of academics involved. Contrast this, instead, with an academic who has never engaged with stakeholders and sees little value in it (the tension between the theoretical and "applied" among researchers) being told they have to do it because it is now part of research assessment. It can come as no surprise that as a result of this approach, there is considerable resistance. Yet, and we reiterate that we are not saying this should be part of habitus for all academics, for some academics the frustration comes from a different perspective – that of management interference as soon as the policy space decided it was important. None of those academics we have spoken with "did impact" because they were motivated to by the rules of REF2014 or REF2021. In a number of cases, they have been doing it for a lot longer than that, in others these academics had deep and mutually beneficial relationships with their stakeholders. The last thing they needed for motivation was a zombie manager repeating lines about engaging with stakeholders, evidencing impact, questioning their underpinning research or counting testimonials, and measuring their significance, when they could already do this. They were doing impactful research as part of their research identity; the REF and the case study demands that come with research assessment were not the motivating factor.

In this chapter we adopted a theoretical stance on academic identity, largely informed by Bourdieu's work in this area. The neoliberal motivations of 21st century higher education policy clearly very strongly conflict with Bourdieu's views on academic motivations around prestige, and the habitus of academia. This is not something that can be imposed, it has to be recognised as a form of prestige capital by academics, and be part of their own habitus.

By recentring the discourse on the alignment between research impact and academic identity, the text advocates for a future where universities not only measure impact through external metrics but cultivate it through the passionate engagement of their academic communities, thereby fulfilling their civic missions in profound and lasting ways.

In conclusion, the integration of research impact into academic identity is multifaceted and complex. Academics often place more faith in their disciplinary peers than in institutional directives, leading to resistance against management-driven impact initiatives. Those who naturally incorporate impact into their work do so out of personal conviction and intrinsic motivation, viewing it as essential to their research rather than an imposed requirement. This perspective aligns with Bourdieu's concepts of field, habitus, and prestige, where academic identity is shaped by deeply ingrained dispositions and the pursuit of scholarly recognition.

Exploring the neoliberal influences on higher education reveals a significant tension between traditional academic values and market-driven policies. Neoliberal approaches often prioritise measurable outcomes, efficiency, and short-term impacts, which can conflict with the long-term, intrinsically motivated nature of impactful research. This creates a challenging environment for academics who value intellectual freedom and the pursuit of knowledge for its own sake, leading to a disconnect between institutional goals and individual motivations.

To foster a genuine commitment to research impact, it is crucial for institutions to support and motivate academics by recognising the intrinsic value of their work and providing a conducive environment for meaningful engagement. This involves aligning impact initiatives with the core academic values of curiosity, critical thinking, and societal contribution. By doing so, institutions can bridge the gap between policy demands and academic identity, ensuring that impact becomes a respected and integral part of scholarly practice.

Ultimately, embedding impact within academic identity requires a shift from top-down impositions to a more supportive and inclusive approach. Recognising and valuing the personal and professional motivations of academics can lead to a more engaged and impactful academic community. This alignment can help universities fulfil their civic missions and contribute positively to society, demonstrating the true value of research beyond traditional metrics and assessments.

References

Blackmore, P., & Kandiko, C. B. (2011). Motivation in academic life: A prestige economy. *Research in Post-compulsory Education*, *16*(4), 399–411.

Bourdieu, P. (1988). *Homo academicus*. Stanford University Press.

Chubb, J., & Watermeyer, R. (2017). Artifice or integrity in the marketization of research impact? Investigating the moral economy of (pathways to) impact statements within research funding proposals in the UK and Australia. *Studies in Higher Education*, *42*(12), 2360–2372.

Fisher, D. (1990). Homo academicus. *Journal of Higher Education*, *61*(5), 581–591.

Fyfe, A., Coate, K., Curry, S., Lawson, S., Moxham, N., & Røstvik, C. M. (2017). *Untangling academic publishing: A history of the relationship between commercial interests, academic prestige and the circulation of research*. St Andrews University.

Kwiek, M. (2021). The prestige economy of higher education journals: A quantitative approach. *Higher Education*, *81*(3), 493–519.

Skea, C. (2021). Emerging neoliberal academic identities: Looking beyond Homo economicus. *Studies in Philosophy and Education*, *40*(4), 399–414.

Śliwa, M., & Kellard, N. (2021). *The research impact agenda: Navigating the impact of impact*. Routledge.

Taksa, L., & Kalfa, S. (2015). Illusio in the game of higher education. *Pierre Bourdieu, Organization, and Management*, *162*, 162–186.

Thomson, P. (2014). Field. In Grenfell, M. (Ed.), *Pierre Bourdieu* (pp. 65–80). Routledge.

Wacquant, L. J. (1989). For a socio-analysis of intellectuals: On "Homo Academicus". *Berkeley Journal of Sociology*, *34*, 1–29.

Watermeyer, R. (2014). Issues in the articulation of 'impact': The responses of UK academics to 'impact' as a new measure of research assessment. *Studies in Higher Education*, *39*(2), 359–377.

Williams, K. (2020). Playing the fields: Theorizing research impact and its assessment. *Research Evaluation*, *29*(2), 191–202.

8 The Impact Ecosystem and Can We Fix It?

Introduction

In the previous four chapters we have conducted a detailed empirical exploration of the issues around research impact in order to answer the question of what research impact is and, most importantly, why it matters. We have also attempted to understand the barriers to greater understanding of impact using the lens of Bourdieu to frame academic identity as a crucial element in moving impact from, as our considerable empirical work shows, a shallow appreciation of impact and institutional views, that it is something that is done because that is what research assessment frameworks call for, to a deep embedded understanding.

In drawing everything together, we will review the discussions and evidence presented across the different chapters and draw the threads together around public good, research policy, research practice, and the evidence base for what research impact currently looks like and if it is fulfilling its intended policy goals. As we unpick the findings of this work we move to propose a new theory around the nature of deep impact culture, driven by an understanding of the public good and how research findings form part of the societal role of universities which draws both upon ethnographic experiences over the last 15 years, policy analysis, and empirical work with peers, that evidences that highly rated impact case studies will generally result from impact being a core aspect of the research activity, and have academics at its core, rather than something that is thought about after the research activities in a project have been conducted. The identity around the impactful researcher, as a part of core academic identity, is positioned as a fundamentally important part of effective research culture, and the resultant public good, and it is important to see the impact ecosystem as a holistic view where multiple stakeholders all work for the same goals.

Bringing Things Together – A Holistic View of Impact

We have argued in this book that an impact strategy that is focused almost entirely on research evaluation will be doomed to focus upon shallow

DOI: 10.4324/9781003454991-8

impact that can be easily packaged in a 5–7-year cycle, and calls for a greater understanding of the roots of the civic university and the role research impact has in placing higher education institutions at the heart of their communities and society, rather than the isolated ivory tower on the horizon.

The survey of Freedom of Information (FOI) data collected from 108 UK institutions reveals a diverse and often inconsistent approach to research impact, highlighting a heavy reliance on external consultants and process. This dependence suggests that institutions may lack confidence in their own capacity to manage and articulate impact effectively or believe that it is something that can be done if you follow a process and then externally validate it. The ad hoc nature of these practices, coupled with the absence of robust policies for selecting and developing impact case studies or support for academics that have demonstrated they have impact, points to a broader issue within the sector around the motivations for doing impact. The inconsistent application of policy and practice underscores the need for a more systematic approach to embedding research impact into institutional frameworks.

Training for research impact is generally provided across institutions, but the nature and quality of this training vary widely. Most institutions offer training through professional services or external consultants, rather than utilising the expertise of impactful academics. Therefore, the view being communicated to academics, the ones who ultimately conduct the research work that is transformed into impact case studies, is not an aspect of academic identity but whose motivations are external and carried out "because we have to for the REF". This method of delivery indicates a disconnect between those who perform impactful research and those who train others, potentially leading to a, arguably, superficial understanding of impact within the institutions themselves. The emphasis on professional service staff to deliver training, without adequate training for themselves, further complicates efforts to build a cohesive impact culture within institutions or for a deep knowledge. Unless impact professionals have taken it on a deeper interest in impact themselves, this "train the trainer" model simply perpetuates shallow understanding of impact ideology and the view that this is a process which, if followed, results in impact.

The analysis also sheds light on workload allocations and the recognition of research impact in academic roles. Only a few institutions have specific workload allocations for impact, and many academics report that impact activities are not considered in their appraisals. This lack of formal recognition and support for impact work will obviously discourage academics from pursuing impactful research, perpetuating the view that impact is a secondary concern imposed by management rather than a core element of academic identity. The absence of consistent policies for selecting impact case studies further exacerbates this issue, making it challenging to integrate impact into academic practice effectively and does not fill those whose careers might depend on successful research assessment outcomes with confidence that case studies are selected based upon a deep understanding of what impact is and how it is assessed.

The heavy reliance on external consultants for research impact during REF 2021, with institutions spending significant amounts on these services, highlights a critical issue in the sector. This reliance indicates a perception among senior leaders that impact can be outsourced rather than developed internally. The view that impact is a commodity that can be bought, rather than a practice that should be nurtured within the academic community, reflects a deeper misunderstanding of the nature of research impact. The survey results suggest that for research impact to be truly embedded in academic culture, institutions need to invest in building internal expertise and recognise the value of impact work within their academic frameworks.

Interviews with impact professionals provide a detailed analysis of their experiences and perspectives in the field of research impact. It highlights how individuals often transition into impact roles serendipitously, driven by a desire for more meaningful work and the recognition of gaps in their current roles. This shift is frequently facilitated by institutional changes and external opportunities, such as the introduction of frameworks like the Research Excellence Framework (REF) in the United Kingdom, which prompted institutions to create new roles focused on impact.

The interviews helped us understand the ethical landscape of the impact industry, emphasising the need for genuine, long-term benefits over superficial metrics. It addresses the challenges within academia, including the undervaluation of applied research and the struggle to balance academic responsibilities with impact activities. The narrative underscores the importance of support structures and institutional backing in fostering a robust impact culture. Effective leadership, strategic vision, and dedicated roles are critical in embedding impact into academic practice and ensuring that impact activities are well-coordinated and effectively executed.

As with the analysis of FOI data, the interviews provided further evidence of the broader institutional struggles to fully support impact and integrate it into the research culture. There is a notable disconnect between impact as an academic pursuit and its perception as a managerial imposition. Among impact professionals there were calls for a deeper integration of impact into research culture, recognising the need for a cultural shift within academia to appreciate the value and importance of research that directly benefits society. And there were views that this involves changing the way we approach impact, creating clear pathways for career advancement, providing professional development opportunities, and fostering a culture that values diverse contributions.

These views were further reinforced with our interviews with academics, all of whom, we should stress, have been effective in developing research identities that are highly impactful (even if, in one case, this was not formally recognised by either the academic or their previous institutions). In exploring the barriers academics face in engaging with research impact, a significant issue highlighted is the perception that impact activities are an add-on rather than an integral part of academic work. This stems from a top-down approach where impact is imposed by university management without adequate support or integration into the academic culture. Academics often feel that impact activities are driven by administrative goals, particularly the Research Excellence Framework (REF), rather than being aligned with their research interests and intrinsic motivations. And there was a view among all of those we interviewed that their institutions and senior managers had little appreciation of impact as being an important part of research culture. They were imposing impact practices upon academics because there was money associated with it. Of course, this was very much in tension with these academics, who were all doing impact because they thought it was part of their research identity, and often did it in spite of the direction taken by their institution, instruction from research and innovation departments, or senior management rhetoric. This disconnect results in a superficial engagement with impact practices, where activities are undertaken to satisfy institutional requirements rather than to achieve genuine societal benefits.

Another critical barrier is the lack of institutional support and training tailored to developing meaningful impact. The interviewees pointed out that while many institutions offer impact training, it is often delivered by external consultants who may not fully understand the intricacies of the specific academic environment. Confirming our own views drawn from the FOI data, interviewees felt that this "train the trainer" model leads to a shallow understanding of impact, focusing on compliance and process rather than fostering a deep, strategic integration of impact into research activities. Academics reported that the training often lacks depth and fails to address the practical challenges they face in achieving real-world impact – they are simply told a process to follow and, as if by magic, a case study will emerge. Consequently, there is a need for more robust, context-specific training that genuinely supports academics in developing and implementing impactful research.

Furthermore, the interviewees discussed the broader cultural and structural issues within academia that hinder impactful research. These include the underappreciation of applied research (even if that research produced demonstrably high-quality outputs), the dominance of traditional academic outputs in performance evaluations, and a lack of recognition for impact activities in career progression. Academics often face heavy workloads and time constraints that leave little room for engaging in impact work. Additionally, there is a gendered dimension to the undervaluation of impact roles, with female academics potentially disproportionately moving into these positions, which are often seen as less prestigious compared with traditional research roles. The document calls for a cultural shift within institutions to recognise and reward impactful research, integrate it into academic identities, and provide the necessary support and resources to foster genuine, long-term impact.

It would seem to us, and the view of many of our interviewees, that the process of "enabling impact" for research assessment purposes is a fundamental barrier to deep, embedded, related impact as part of research culture. And the key issue here lies in the almost obsessive focus of the case study as the goal of impact activity.

We are reminded, as frequently, of the comment from our former DVC colleague, who, as discussed in Chapter 6, was fond of stating:

> If you do good research and have good impact, research assessment is simply something that comes around, not something you should focus on.

The current state of the sector, we would suggest, is the antithesis of this. Arguably, for the majority of the sector impact occurs not because it is a natural part of the research process and can be written up when research assessment comes around, but because you have to do it for the REF.

When considering the case study as the vehicle for impact, it becomes a reductionist view on what impact can be, and results in a focus on process, not research culture or identity. In the view of impact in the previous quotation, and one that is reflected by many of those we interviewed (and certainly in our own experiences), academics who believe impact is part of their research culture will do it as a natural part of their overall research identity, and when research assessment emerges, that have outputs and evidence in abundance, and testimonials are easily accessed because their stakeholder relationships are strong.

If the starting point for impactful research is "you need to write a case study" it will only ever be a process, beset by "mock refs" where draft case studies (often many years prior to submission) are assessed, in a lot of cases by people without a good knowledge or research, and an army of peers and managers tell the academic under assessment to "do more", often with little support. We are reminded of the colloquialism "you do not get a pig fat by weighing it". One does not nurture impact as part of research culture by demanding academics spend time loading evidence and testimonials into a system "so we can monitor progress" or telling them to spend their summers writing draft impact case studies. Nurturing comes for support, discussion, and recognising that those academics that do it well will do it because they believe in it, not because they have been told to do it.

There is a need to break away from the case study as the driver for impact. Of course it is an essential part of the research assessment exercise and therefore important, but having a case study as a starting point reduces the impact approach to one of simplistic repetitive tropes such as:

What is the underpinning research?
Who are the stakeholders?
What is the underpinning evidence?
What is significance and reach?

And therefore, you have done impact.

All of these questions are important, and indeed we have found ourselves asking these to colleagues over the years. However, an impactful academic with deep knowledge of why they are doing it, with excellent stakeholder relations where everyone works for mutual benefit, will have all of these answers to hand, when there is a need to write up their impact. And those who do not, will not gain a deep understanding of impact from being told reductionist tropes such as "You need to work with more stakeholders".

There have been previous authors who have argued that impact is at odds with academic identity and research culture. Certainly, these are very articulate arguments and we would not disagree with the overall premise that impact, as imposed by policymakers with "value-for-money" arguments suggesting that if academics are not doing research that has commercial, social, or cultural value, and senior managers who are demanding it is done not because they think it is important, but because they want to maximise their REF return, this approach to impact is certainly at odds with academic identity.

However, as we have discussed in this book, we have spoken with a number of academics with a great deal of impact success, in REF terms, but who also feel their approach is done in spite of, rather than with the support of, policymakers and leadership. They are doing this because they see it as a part of their academic identity, and they enjoy it. As with the vignettes presented in Chapter 1, they were doing impactful research because it mattered to them. The policy space just played catch up and suddenly the managers who had previously told them not to waste their time disseminating their research outside of core academic outputs became interested in this "impact stuff".

Equally, we have spoken to impact professionals who can see the flaws in the system, and how a process- and measurement-obsessed approach to impact is sullying the quality of what is being carried out, often, independently by academics who were doing it regardless of their institutions. However, these professionals often felt constrained by the system and did not feel that they could break out of this model, because of frustrations from impact managers who recognise the flaw in the system but don't feel that they can break the model because that's what external people are telling their "superiors" (in organisational terms) works, and the superiors do not have sufficient knowledge (or indeed motivation) to challenge this view.

In terms of the most effective relationship between impact professionals and academics, we reached something of a consensus from both sets of subjects with whom we interviewed. And this also reflects with our own experiences in working together to achieve high-quality impact (and indeed created one of the motivations for doing this book).

Essentially, it is not the job of the impact professional to tell the academic what to do, it's a case of understanding their research and seeing how they might be supported with, for example, relating their work to impact approaches, or to make suggestions on how they might improve reach or significance.

From both interviewees and in our own experience, the best things are support and listening/providing a sounding board (and acknowledging that knowledgeable impact managers know a lot about research policy and how it might be applied to specific research practices). It is not demanding they write draft case studies that are then read by other people who don't understand impact and tell them to do more/publish more/get more testimonials/engage with more stakeholders which are in essence all repeated tropes which have little understanding of the research process or academic motivation and instead viewed the academic as a cog in the machine rather than the engine.

Reflecting on the working practices of the two authors of this book, we have already touched upon the value in having a sounding board with a deep understanding of the research policy space, but more importantly knowledge of how academics conduct research. This is, to some degree, as a result of Louise having a PhD and having worked in academic roles before becoming an impact professional. Ultimately, the work we have done in preparing the case study for REF 2021 is a co-creation. However, there is also a stronger working relationship that helps the long-term goals of Andy's research and we continue to work together for improved mutual practice even though we now work in different institutions.

Are we saying, therefore, that impact professionals need to have a PhD? No, this would be prescriptive and that generally does little to nurture a harmonious ecosystem. But having knowledge and understanding of academic custom, and identity, is important, as is an appreciation that impact is not a measure of research assessment, it is, for some academics, at the very heart of their practice. What we need less of is what we have referred to in this text as zombie staff, who repeat what they have seen in presentations, with an eye on institutional policy and the expectations of senior management rather than academic motivation.

The other thing that comes out strongly from these interviews, and something in conflict with some of our discussion with some institutions about their approach to research culture, is a simple one that is often, it seems, overlooked.

Not everyone needs to be doing impact.

It seems to us, as participants in higher education for many years, that a poorly understood policy influence or demand often results in a knee jerk, irrational response by institutions. We have seen institutions mandate that everyone needs to do an element of impact (or engagement) within their own practice, just because it is now assessed. And, equally, we have had many conversations over the years with academics with excellent profiles, and whose outputs are of a very high quality, suddenly trying to define their work within an impact case study as if this is now a measure of research excellence in the institution, which invariably involves a retro-fitting of findings to potential stakeholders or attempting to build stakeholder relationships with no foundation in an unconvincing manner, they are not doing it because they believe it in or are passionate about it, but because they think they should for whichever research assessment exercise is looming.

Certainly, there are some disciplines and research directions that are far better suited to impactful practice than others. And there are areas of blue-sky research whose "impact" might not be felt for many years, or not at all. And that should be fine. We still need this spectrum of research practice and just as some of our interviewees felt that their applied practice was viewed as second class by some academic leaders in their institutions, in other places those who did not have clear routes for impact felt like they had to.

It is clear from our interviews that impact can be part of academic identity and doesn't have to be viewed as an imposition. However, that does not mean that everyone who is an academic must do impact or have it as part of their academic identity.

The Research Impact Ecosystem

When we started this project, we were a little concerned that we might be the outliers and either the sector had impact well-embedded, or academics understood its value and were enthused by its addition in research assessment.

However, we must conclude, having gone through the various aspects of this project, that the state of the sector is actually far worse than we imagined, and while we met plenty of kindred spirits whose enthusiasm for impact came from their own engagement in their field, and who were doing it because they believed it was important, often explicitly against the advice of their line managements and senior leaders, there were others who were frustrated by the lack of support, poor advice received from management and consultants, and an annoyance at a model of research management at their institution that seemed process-focused with familiar impact tropes being delivered to instruct the academic how to achieve the case study desired by the institution for their REF submission. Or, alternatively, being told by the institution not to worry about impact because "we've got enough case studies now".

Whose fault is this?

While Bourdieu's theories on academic identity have been extremely useful to understand how, intrinsically, a better model of impact might be developed in line with how academics work and how they are motivated, the concepts of the field, capital, prestige autonomy, and habitus make a framework that is internalised around academia, and academic identity, and while, in our examination of the concepts in Chapter 6, those academics who see impact as part of their identity would also view their field and habitus as reaching beyond academia, as it is fundamentally a theory of academia.

There is also a need for a macro view of the impact ecosystem as a whole to help better understand how it should operate in a holistic manner, and also in the expanse of the research impact space. What seems clear from our empirical work in this book is that there is rarely a macro view of the impact space, more a disjointed model where pockets of knowledge do not become institutionally dominant and where academics are viewed as subordinate.

Fundamental to all of our interviews, and also our own experiences over many years working in higher education, is that deep impact starts with engaged and motivated academics. Positioning academics in the stakeholder model highlights their relationship to the field and removes the idea of them being outside of it. Therefore, we propose that ecological systems theory (EST), which was explored in a domain specific way in Chapter 2, can equally be applied in this context.

As already discussed, but worthy of review in this context, Bronfenbrenner's ecological systems theory (Bronfenbrenner 1979) is a framework for understanding human development within the context of the system of relationships that form their environment. It was developed by Urie Bronfenbrenner, a developmental psychologist, who emphasised the importance of studying children in multiple environments, also known as ecological systems, in the attempt to understand their development.

The theory outlines five environmental systems that interact to influence human development:

- Microsystem: This is the closest layer to the individual and includes the structures with which the individual has direct contact. It encompasses relationships and interactions a person has with their immediate surroundings, in the original this would include actors such as family, school, peers, and neighbourhood. At this level, interactions are bidirectional, meaning the individual and the environment influence each other.
- Mesosystem: This layer provides the connection between the structures of the individual's microsystem. In Bronfenbrenner's original work these are the relationship between a child's family and their school, or between their peers and their family. The mesosystem focuses on how these relationships interact and affect the individual.
- Exosystem: This system includes the broader social systems that do not directly contain the individual but affect them indirectly. For child development this would include a parent's workplace, community services, or neighbourhood safety. While the individual does not interact with the exosystem directly, they feel the positive or negative force involved with the interaction with their own system.
- Macrosystem: This layer encompasses the cultural context, including societal values, customs, and laws. The macrosystem influences how the systems within the smaller layers function. For example, cultural values may dictate the resources available for child development, societal norms, and even the structure of educational institutions.
- Chronosystem: This system encompasses the dimension of time as it relates to an individual's environments. Elements within this system can be either external, such as the timing of a parent's death, or internal, such as the physiological changes that occur with aging. The chronosystem involves the patterning of environmental events and transitions over the life course.

Bronfenbrenner's theory highlights the importance of considering multiple levels of influence on an individual's development and recognises that interactions between these systems can be as significant as the influences themselves. This ecological approach to understanding development has been influential in fields such as social work (Siporin 1980), education (Bluteau, Clouder and Cureton 2017), youth services (Duerden and Witt 2010), and public health (Crawford 2020).

In higher education, it has been used to understand how various levels of influence, from institutional policies to institutional practices (Kitchen et al. 2019), investigating how interactions between students, peers, faculty, and institutional structures, impact academic performance and well-being (Renn and Arnold 2003; Zwemer et al. 2022), understanding how ecological factors affect faculty job satisfaction, productivity, and career progression (McLaughlin 2020), and examining how external policies, economic conditions, and cultural contexts influence academic institutions' operations and strategies (Ross et al. 2022; Chan, Mazzucchelli and Rees 2021).

The literature first helps to show how adaptable the model is in different contexts, a predominant factor in all of these applications is a demonstration of the model's utility in addressing multifaceted issues within the academic ecosystems and helping to inform holistic and effective educational strategies and policies. As such, its application to helping understand a holistic view of the research impact ecosystem seems appropriate, and a model that can place the academic at the centre of the model.

This is not, however, a complete application of the whole model to the context. For example, with a lack of longitudinal data it is difficult to observe change over time or the developmental aspects of an impactful academic. However, as a tool to help us understand why research impact remains poorly understood and what we might do about it, the EST approach provides a model for us to understand the ecosystem, in the same way that Bourdieu's model helped us understand why academics often do not engage with the concept of impact. And, we would propose, a more holistic view of the research impact ecosystem is a very useful tool to articulate the challenges of achieving deep impact knowledge within an organisation and how the wider ecosystem needs to react. Taking the core ecosystem model as a foundation, we can map different levels of stakeholders and influence onto the macro-, exo-, and microsystems (Figure 8.1).

In exploring the stakeholders in the systems, we will initially present an ideal ecosystem model, prior to presenting an analysis where, according to our data, things are not working as well as they might.

Microsystems

- **Researchers/peers:** Those academic peers who might work directly with the academic on research projects but might also contribute to the wider body of knowledge which informs the academic's own research practice.
- **Academic leaders:** Those who are involved in direct interactions that researchers have in their immediate environment. This includes relationships and collaborations with peers, mentors, students, and

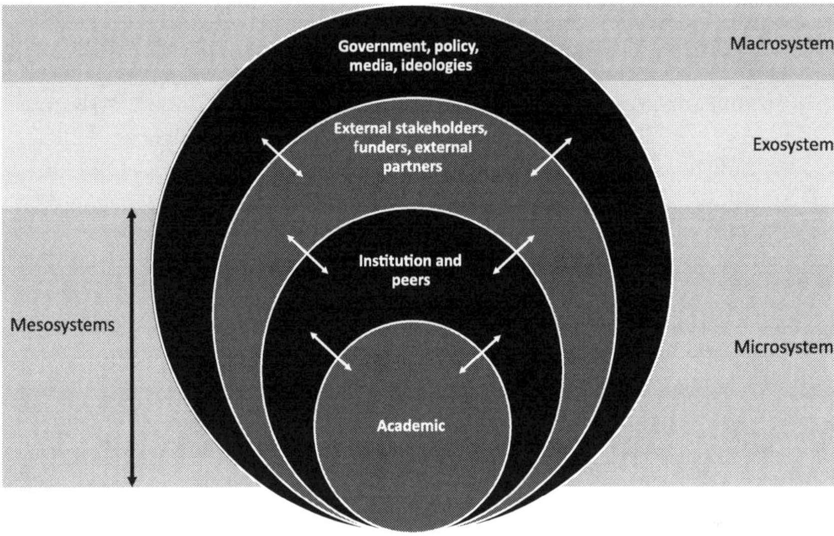

Figure 8.1 The research impact ecosystem.

immediate departmental colleagues. Institutional management should provide clear policy on impact and how the institution implements impact strategies.

- **Impact professionals:** Helping inform policy in the institution based upon knowledge of the research policy space more widely, identify funding opportunities and occasions for collaboration and providing support for academics in how their research might be better articulated in an impactful manner.
- **Academic institution:** The academic's employer who broadly hosts and supports researchers, provides infrastructure, and facilitates collaboration. At an institutional level there should be senior leadership who develops and implements impact strategy that is mindful of the research and impact strengths of the academics employed.

Mesosystem

- Mesosystems in the application of EST to the research impact ecosystem play a pivotal role in enhancing research impact by creating and maintaining interconnected networks that facilitate various supportive and collaborative interactions. They can link different departments and disciplines within an academic institution. Additionally, mesosystems should include the relationships between institutions and external research entities such as other universities, research institutes, and professional organisations. Furthermore, mesosystems encompass the crucial interactions between researchers and external stakeholders, such as industry partners, governmental agencies, and non-profit organisations. These relationships are essential for translating research findings into practical applications and policy recommendations, thereby maximising research impact. Within the institution, mesosystems include interactions among various support units, such as research administration, grant offices, and technology transfer offices, ensuring researchers receive the administrative and financial support needed for their work.

Exosystem

The exosystem is, essentially, external stakeholders, funders and other external partners and comprise actors such as:

- **Industry and business:** Applying research findings to develop new technologies, products, and services.
- **Non-governmental organisations (NGOs) and non-profit organisations:** Who can make use of research to support advocacy, programme development, and interventions and inform practice with research findings to, for example, address societal issues and support vulnerable populations.
- **Community groups:** Engaging with research that affects their community and also to participate in research projects if appropriate. Community groups can ensure research addresses community needs, improves local conditions, and fosters community engagement.
- **Policymakers (as impact stakeholders):** Making use of research findings to inform and develop policies, regulations, and laws, and to ensure that policy development is evidence-based.
- **Funders:** Setting funding calls and making clear statements on the needs for impact in funding applications, articulating key stakeholders, and embedding impact into the funding process (if appropriate).
- **Consultants:** Consultants also fit into the exosystem and in an ideal ecosystem would support the work of academics in identifying research strengths and impact potential, and facilitate the understanding of impact by those academics.

Macrosystem

- **Policymakers (setting impact policy):** Policymakers who set research policy shape the overarching policies, regulations, and funding priorities that influence the research environment. Policymakers can create frameworks and guidelines that encourage or mandate research impact, setting the tone and expectations for how research should contribute to societal needs and priorities. For instance, as we have discussed at length in this book, the Research Excellence Framework (REF) transformed institutional interest in research impact, although perhaps not in a holistic or embedded manner. In Bronfenbrenner's model, there is an expectation that there is two-way communication between systems and across systems. A crucial aspect of this at the macrosystem level is the need for policymakers to acknowledge their role in driving metric-based, process-focussed approaches to research assessment, and the impact of these assessment frameworks on academic practice. And we are not convinced this is always recognised at present.

- **Media and communications:** While not directly in contact with the research and impact process, media can be a useful stakeholder in disseminating research findings to a broader audience to inform the public, raise awareness, and facilitate public discourse on academic research.
- **Regulatory bodies:** Oversee compliance with laws and regulations that research may impact, ensuring public safety, ethical conduct in research, and maintaining standards. In the UK, the predominant regulatory body is UK Research and Innovation, who define the REF.
- **Ideology:** Ideology within the macrosystem is crucial in shaping norms, values, and policies governing academic research. Ideological beliefs about the purpose and value of research drive societal expectations and priorities. Understanding academic ideology (as explored in Chapter 6) is fundamental to a healthy ecosystem, because imposing policy that does not align with ideology will always cause friction in the system.

Thus, we described what would be an ideal ecosystem. We would propose that, at the present time, as informed by our research, the ecosystem does not necessarily reflect this ideal, and we illustrate that in Figure 8.2.

It is perhaps not fair to say that the ecosystem around research impact is broken, but there is considerable evidence to suggest it is certainly not working in an optimal manner and a lot of impact-based outcomes rely on consultants and process rather than deep knowledge of impact and motivated (and supported) academics within institutions. Perhaps dysfunctional rather than broken.

As we have discussed throughout previous chapters, and summarised in this modelling, we would suggest that the fundamental issue within the ecosystem is one of deep knowledge across the different systems. The fundamental implication of a dearth of knowledge in the microsystems is first the academic will receive poor support and become demotivated by senior managers who instruct approaches that the knowledgeable academic will already recognise as flawed. The breakdown of communication in the mesosystems can further break down the notion of microsystems working together for mutual goal, and weakened senior management leads, we are told, to the recruitment of subordinates whose knowledge is also weak (because a senior leader who lacks knowledge does not want to feel undermined by a subordinate who knows more than they do). If the microsystems are not coherent and strong, the influence of external actors and policymakers becomes greater

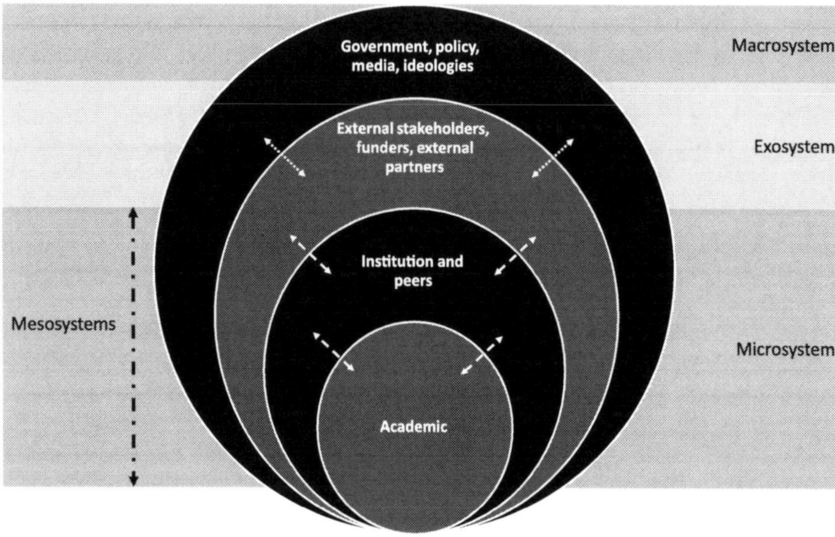

Figure 8.2 A dysfunctional ecosystem.

than should be the case in a working ecosystem, and they will drive process models and "solutions" into institutions and, if the senior managers know no better, result in the delivery of the impact agenda without any institutional understanding of impact. If communication upwards from microsystems to policymakers is not effective, institutional knowledge is not strong, and there will be too strong a push from policymakers in terms of what they believe what impact should look like, rather than what academics can most align with and apply to their research identity. And the stakeholders in the exosystem who could benefit from impactful research will not achieve the best outcomes because the impact articulation is aligned with policy demands, not the public good.

So, what would it take to fix?

It is clear from our interviews that the issues lie in organisational understanding and knowledge of impact, and how to support it. We have explored this at length in earlier chapters and we would suggest a strong steer from our interviewees is that the core of the most effective knowledge, the academic community, is, in many cases, ignored, with faith placed instead in external consultants who have "the solution". Does this mean that we think that consultants should not have a role to play in research impact? Of course not, we have worked with excellent consultants who add a great deal of value in being able to bring cross-institutional knowledge and excellent personalised advice for academics. However, we have also seen consultants who purport to have THE solution, and if the institution commissions them (at considerable cost if the analysis from the FOI work is a good indication), they will guarantee excellent results in the next research assessment.

Impact literacy is not solely about understanding the metrics of research assessment, it is about a deep understanding of research, and how research findings can transform society. At its core, impact has to come from the academic community, and it has to be part of their identity. We have spoken to academics who are knowledgeable about how their work can have impact and have strong, long-term relationships with stakeholders. They generate excellent impact because impact is part of their research process, it is not a bolt on or something that might be done when the funding runs out and the publications have been done. Any imposition of impact policy that does not align with this will immediately cause friction and lack of buy in with the academic community and, as a result, impact will never become embedded institutions. And while impact professionals can sometimes be viewed by some academics as facilitators of the neoliberalisation of higher education and the devaluing of academic autonomy, this was not the case with those with whom we spoke, who could both see the flaws in the current system and also appreciate how impact professionals and academics can work harmoniously to build deeper impact in academic practice that will result in higher-quality impact for assessment exercises but not be driven by them.

Conclusions

In conclusion, the exploration of research impact within the context of higher education reveals several critical insights and challenges. Throughout the previous chapters, we have examined the multifaceted nature of research impact, emphasising its significance not just as a metric for research assessment but as a fundamental aspect of academic identity and practice. Drawing on empirical data, including Freedom of Information (FOI) responses from 108 UK institutions, it is clear that the approach to research impact is inconsistent and often reliant on external consultants. This reliance suggests a lack of internal confidence and capacity to manage and articulate impact effectively, highlighting the need for a more integrated and institutionally embedded approach.

Our analysis of the FOI data showed a significant variation in how institutions support and recognise research impact. Training programmes for impact are widespread but appear to vary greatly in quality and delivery methods. Many institutions depend on professional services or external consultants rather than leveraging the expertise of impactful academics, which can lead to a superficial understanding of impact among researchers. This disconnect between those who perform impactful research and those who deliver training perpetuates a view of impact as a compliance exercise rather than a core academic endeavour. To foster a deeper impact culture, institutions must integrate impact training into the academic fabric, utilising the knowledge and experience of successful impact-driven researchers.

Interviews with impact professionals further underscore the systemic issues within academia regarding research impact. Many professionals expressed frustration with the process-oriented approach to impact, which often overlooks the intrinsic motivations and expertise of academics. The interviews revealed that impactful academics engage in this work not because of institutional directives but because they believe in its importance and relevance to their research. To address this, institutions need to shift from a top-down imposition of impact policies to a more supportive and collaborative environment that recognises and nurtures the intrinsic motivations of researchers.

So, to return to the question at the start of this book. Why is impact important?

From the perspective of the interviewees, and our own view, it is because it can result in real change or benefit to stakeholders and wider society, that is informed by the sort of research that academics are passionate about. Or, to directly quote one of our interviewees, whose work has had significant impact upon the broader field around their discipline, from grass roots to government, over many years:

> Because I enjoy it, and I'm good at it.

This is not someone who is engaged in impact work because they see it as a useful alternative to writing outputs. Indeed, they are incredibly well-cited in their field and frequently publish high-quality outputs. But with their research discipline, around digital education broadly with a specific focus on science, technology, engineering, and mathematics and curriculum and policy development, engaging with stakeholders at all levels of the policy space, it is important and ensures research findings can transform policy.

For them, it is a fundamental part of their academic identity, and this was the same for all of the academics with whom we spoke who were already engaged in impact framed within research assessment. However, their motivation was not research assessment, it was seeing the value of their work to wider society.

And because, for others, it could be part of their identity if it's articulate in a way that is relatable to them. And the way to approach that is certainly not to tell them "you have to do this because the institution will get more money (and by the way, you probably won't see any of it)".

Ultimately, for research impact to be truly embedded within academic culture, a comprehensive and cohesive approach is necessary. This involves clear policies, adequate training, and institutional recognition of impact work as a core component of academic practice. Institutions must invest in building internal expertise, fostering a supportive culture, and integrating impact into career development frameworks. By doing so, they can move beyond a superficial engagement with impact and towards a more meaningful, sustainable, and impactful academic environment, with trusted and respected academics at the centre, not a mere cog in the machine.

References

Bluteau, P., Clouder, L., & Cureton, D. (2017). Developing interprofessional education online: An ecological systems theory analysis. *Journal of Interprofessional Care*, *31*(4), 420–428.

Bronfenbrenner, U. (1979). *The ecology of human development: Experiments by nature and design*. Harvard University Press.

Chan, H., Mazzucchelli, T. G., & Rees, C. S. (2021). The battle-hardened academic: An exploration of the resilience of university academics in the face of ongoing criticism and rejection of their research. *Higher Education Research & Development*, *40*(3), 446–460.

Crawford, M. (2020). Ecological systems theory: Exploring the development of the theoretical framework as conceived by Bronfenbrenner. *Journal of Public Health Issues and Practices*, *4*(2), 170.

Duerden, M. D., & Witt, P. A. (2010). An ecological systems theory perspective on youth programming. *Journal of Park & Recreation Administration*, *28*(2), 108–120.

Kitchen, J. A., Hallett, R. E., Perez, R. J., & Rivera, G. J. (2019). Advancing the use of ecological systems theory in college student research: The ecological systems interview tool. *Journal of College Student Development*, *60*(4), 381–400.

McLaughlin, K. (2020). *Online adjunct faculty members' perspectives on work and nonwork balance: A study of ecological-system influences*. Saint Joseph's University.

Renn, K. A., & Arnold, K. D. (2003). Reconceptualizing research on college student peer culture. *The Journal of Higher Education*, *74*(3), 261–291.

Ross, P. M., Scanes, E., Poronnik, P., Coates, H., & Locke, W. (2022). Understanding STEM academics' responses and resilience to educational reform of academic roles in higher education. *International Journal of STEM Education*, *9*(1), 11.

Siporin, M. (1980). Ecological systems theory in social work. *Journal of Sociology and Social Welfare*, *7*, 507.

Zwemer, E., Chen, F., Dallaghan, G. L. B., Shenvi, C., Wilson, L., Martinelli, S. M., ... & Shaheen, A. (2022). Reinvigorating an academy of medical educators using ecological systems theory. *Cureus*, *14*(1).

Index

Note: Figures are shown in *italics* and tables in **bold** type. Footnotes are referred to by the page number followed by the footnote number e.g. 130n3 refers to note number 3 on page 130.

Made in the USA
Las Vegas, NV
19 September 2026

dfc81b2b-1b78-4ad5-ac5b-84044bae23a6R02